Emotional States

What is the political allure, value and currency of emotions within contemporary cultures of governance? What does it mean to govern more humanely? Since the emergence of an emotional turn in human geography over the last decade, the notion that our emotions matter in understanding an array of social practices, spatial formations and aspects of everyday life is no longer seen as controversial.

This book brings recent developments in emotional geography into dialogue with social policy concerns and contemporary issues of governance. It sets the intellectual scene for research into the geographical dimensions of the emotionalised states of the citizen, policy-maker and public service worker, and highlights new research on the emotional forms of governance that now characterise public life. Drawing on an international range of case studies, the chapters examine issues of regulation, modification, governance and potential manipulation of emotional affects, professional and personal identities, and political technologies. Contributors provide analysis of the role of emotional entanglements in policy strategy, policy implementation, service delivery, citizenship and participation, as well as considering the emotional nature of the research process itself.

This innovative book will be of interest to researchers and students within social policy, human geography, politics and related disciplines.

Eleanor Jupp is a Lecturer in Social Policy at the University of Kent, UK.

Jessica Pykett is a Senior Lecturer in Human Geography at the University of Birmingham, UK.

Fiona M. Smith is a Lecturer in Human Geography at the University of Dundee, UK.

Emotional States

Sites and spaces of
affective governance

**Edited by Eleanor Jupp,
Jessica Pykett and
Fiona M. Smith**

LONDON AND NEW YORK

First published 2017
by Routledge
2 Park Square, Milton Park, Abingdon, Oxon OX14 4RN

and by Routledge
711 Third Avenue, New York, NY 10017

First issued in paperback 2018

Routledge is an imprint of the Taylor & Francis Group, an informa business

British Library Cataloguing in Publication Data
A catalogue record for this book is available from the British Library

Library of Congress Cataloging in Publication Data
Names: Jupp, Eleanor (Research Associate), editor. | Pykett, Jessica, editor. |
Smith, Fiona, 1968 December 10– editor.
Title: Emotional states: sites and spaces of affective governance / edited by
Eleanor Jupp, Jessica Pykett and Fiona M. Smith.
Description: Abingdon, Oxon; New York, NY: Routledge is an imprint of
the Taylor & Francis Group, an Informa Business, [2017] |
Includes bibliographical references and index.
Identifiers: LCCN 2016021539| ISBN 9781472454058 (hbk) |
ISBN 9781315579252 (ebk)
Subjects: LCSH: Policy sciences—Psychological aspects. |
Social policy—Psychological aspects.
Classification: LCC H97 .E59 2017 | DDC 320.01/9—dc23
LC record available at https://lccn.loc.gov/2016021539

ISBN 13: 978-1-138-62416-0 (pbk)
ISBN 13: 978-1-4724-5405-8 (hbk)

Typeset in Times New Roman
by Keystroke, Neville Lodge, Tettenhall, Wolverhampton

Contents

Contributors

Rosie Anderson is a Teaching Fellow at the Global Public Health Unit at the University of Edinburgh, UK. She was previously a post-doctoral researcher with What Works Scotland, also at the University of Edinburgh, where she focused on practical applications of her doctoral work on reason and emotion in policy-making, and at Northumbria University, where she researched aspects of trust and empathy in UK crowdfunding.

Matej Blazek is a Lecturer in Human Geography at Loughborough University, UK. His research concerns vulnerability, marginalisation and agency, with a particular interest in geographies of childhood and insecure migration. He is the author of *Rematerialising children's agency* and the co-editor (with Peter Kraftl) of *Children's emotions in policy and practice*.

Sophie Bowlby is a feminist social geographer with interests in issues of access, care and friendship. She is now retired but continues to do research as a visiting professor at Loughborough University and a visiting research fellow at Reading University.

Donna Marie Brown is a Senior Lecturer in Social Science at Northumbria University, UK. Her research interests span social geography, social policy and criminology, with a focus on social justice and marginalised groups. Areas of particular interest include youth justice, community policing and participatory action research.

John Clarke is Emeritus Professor of Social Policy at the Open University, UK, where he worked for over 30 years. He is also a Visiting Professor at Central European University in Budapest. His research has focused on changing forms of state welfare, especially the politics of managerialism and consumerism. His most recent publications include *Disputing citizenship* (with Kathy Coll, Evelina Dagnino and Catherine Neveu) and *Making policy move: towards a politics of translation and assemblage* (with Dave Bainton, Noémi Lendvai and Paul Stubbs).

Tom Collins completed his PhD in Human Geography at the University of Leeds, UK. His doctoral research examined the cultural geographies of civic pride in his home city of Nottingham, and drew upon interests in urban and emotional geographies. He has presented his research at a range of international conferences, including the International and Interdisciplinary Emotional Geographies Conference in Groningen, the Netherlands.

Rachael Dobson is a Senior Lecturer in Sociology and Criminology at Kingston University, UK, whose interdisciplinary research is informed by critical social policy, sociology, psycho-social studies and critical race- and feminist-informed scholarship. A central starting point for her work is welfare practices and interventions with people who are constructed as both vulnerable and transgressive. This enables her to draw substantive and relational connections between power and agency, welfare institutions, social policy, the state and governance regimes.

Bryony Enright is a Post-doctoral Research Fellow in the Graduate School of Education at the University of Bristol, UK. She completed her PhD in Human Geography at the University of Birmingham in 2013 in which she examined the role of temporary staffing agencies in local labour markets and the experiences of low-skilled agency workers. Currently Bryony works on the AHRC Connected Communities programme alongside Leadership Fellow Keri Facer. Their research analyses the practice and legacy of collaborative and interdisciplinary research and contributes to current debates about the role of the university.

Keri Facer is Professor of Educational and Social Futures at the University of Bristol, Graduate School of Education, UK. She works on rethinking the relationship between formal educational institutions and wider society and is particularly concerned with the sorts of knowledge that may be needed to address contemporary environmental, economic, social and technological changes. Her recent books include: *Learning futures: education, technology and social change* and *The politics of education and technology* with Neil Selwyn. Since 2013, Keri has been Leadership Fellow for the AHRC/RCUK Connected Communities programme (www.connected-communities.org).

Menno Fenger is Associate Professor of Public Administration at Erasmus University Rotterdam, the Netherlands. His research focuses on processes of policy change and institutional development in social policies in comparative perspective. He also publishes on the effectiveness of policy instruments.

Kirsten Forkert is a Lecturer in the School of Media at Birmingham City University, UK. She teaches media theory and research methods. Her research explores cultural labour and austerity politics. She is the author of *Artistic lives* and *Austerity as public mood*.

Elizabeth A. Gagen is a Senior Lecturer in Human Geography at Aberystwyth University, UK. Her work examines the role of psychological knowledge in governing the spaces of childhood and youth. She has explored this relationship through a range of empirical sites, including late nineteenth and early twentieth-century American schools and playgrounds, military recreation during the First World War and contemporary emotional education in the UK. More recently, she has considered what can be rescued from psychological and psychotherapeutic practices that can enable young people to navigate the spaces through which they are governed.

Mark Griffiths is a Postdoctoral Researcher at the University of Oulu, Finland, and holds a PhD from King's College London in Public Policy Studies. His teaching and research focus on global civil society, citizenship and volunteering for development. Mark has published on embodied experience as both an important aspect of governance and a potential site of resistance or transcendence. Current work includes a monograph on the positionality of Western researchers in the 'Global South'.

Louise Holt is a Reader in Human Geography at Loughborough University, UK. Her research interests as a critical social and cultural geographer focus upon exploring how enduring inequalities are reproduced and/or transformed at a variety of intersecting spatial scales and the ways in which everyday, bodily practices in specific spaces/places are connected to, reproduce and can potentially transform broader-scale inequalities.

Rachel Howell is a Lecturer in Sociology/Sustainable Development in the School of Social and Political Science at the University of Edinburgh, UK. Her research interests include lower carbon/sustainable lifestyles, pro-environmental behaviour change, social movements for sustainability, environmental policy and governance, and climate change communication. Prior to that she was a Postdoctoral Research Fellow at Aberystwyth University, funded by the Climate Change Consortium of Wales.

Shona Hunter is Associate Professor in Sociology and Social Policy Governance, University of Leeds, UK, and a Visiting Associate Professor in the Visual Identities in Art and Design Research Centre (VIAD), University of Johannesburg, South Africa. Shona has research interests in all aspects of welfare governance, state practices and the broader material-cultural-affective politics through which the state is enacted nationally and globally. She is the author of *Power, politics and the emotions: impossible governance?* She holds a number of editorial board positions on journals including *Critical Social Policy*, *Policy and Politics* and *Sociology*. She has held visiting positions at the University of Sydney, Australia and the University of Mannheim, Germany.

Emma Jackson is a Lecturer in Sociology at Goldsmiths, UK. Her research, teaching and writing explore everyday practices of belonging and social inequality in cities. Emma is the author of *Young homeless people and urban space: fixed in mobility*, co-author of *The middle classes and the city: a study of Paris and London* and co-editor with Hannah Jones of *Stories of cosmopolitan belonging: emotion and location*.

Hannah Jones is Assistant Professor in Sociology at the University of Warwick, UK. Her research and teaching focus on critical social policy, belonging and inequality, racism and attitudes to migration, and critical and participative research methods. Her publications include *Negotiating cohesion, inequality and change: uncomfortable positions in local government*, which won the 2014 British Sociological Association Philip Abrams Memorial Prize, and *Stories of cosmopolitan belonging: emotion and location*, which she co-edited with Emma Jackson.

Rhys Jones is Professor of Human Geography at Aberystwyth University, UK. His research is situated within the geographies of the state and its related group identities. He is author of a wide range of publications, including *Changing behaviours: on the rise of the psychological state* with Jessica Pykett and Mark Whitehead and *Rescaling the state: devolution and the geographies of economic governance*, with Mark Goodwin and Martin Jones.

Eleanor Jupp is a Lecturer in Social Policy at the University of Kent, UK, and has held research and teaching positions at the Open University, University of Reading and Oxford Brookes University. Her interests straddle social policy and urban social geography, with a focus on marginalised urban neighbourhoods and shifting relations between social policy frameworks and citizens. Areas of particular interest include family policy and 'early intervention', modes of community action within disadvantaged neighbourhoods, activism and austerity, gender and the home within social policy. She has also worked in policy and practice in the voluntary sector, neighbourhood regeneration and social exclusion.

Wendy Larner is Provost of Victoria University of Wellington, New Zealand, and prior to that was Professor of Human Geography at the University of Bristol, UK. Her research is situated in the fields of globalisation, governance, neoliberalism and gender. Recent publications include a co-authored monograph called *Fashioning globalisation: New Zealand design, working women and the cultural economy* (with Maureen Molloy) and co-edited special issues of *Feminist Theory* (Life, ecology, bodies: new materialism and feminisms) and *Social Politics* (New times, new spaces: gendered transformations of economy, governance and citizenship, 2013).

Jennifer Lea is a Lecturer in Human Geography at the University of Exeter, UK. She has research interests in embodied practices, spiritualities and the production and expression of differences related to disability and mental health. She is currently working on a project about parenting and post-natal depression.

Rachel Lilley specialises in teaching mindfulness in the context of the workplace, corporate social responsibility (CSR), leadership and sustainability. She has worked in project management in sustainability for the last 20 years and is a strategist, trainer, facilitator and business coach. She has a research MPhil in mindfulness, pro-climate behaviour and behaviour change from Aberystwyth

University. She works as a specialist consultant on behaviour change and community engagement and was Research Associate on an ESRC-funded behaviour change and mindfulness project. She has previously worked in communications and marketing and as a journalist.

Janet Newman is an Emeritus Professor of Social Policy and Criminology at the Open University, UK. Her research centres on critical analyses of governance, policy and politics, and she brings perspectives from feminism and cultural studies to her research. She is the author or editor of 12 books and has published across a wide range of disciplinary journals. A particular interest is in the interface between activist movements and changing governance regimes: *Working the spaces of power: activism, neoliberalism and gendered labour*. She is also currently exploring the ways in which 'austerity' governance is remaking both politics and culture.

Jessica Pykett is a social and political geographer at the University of Birmingham, UK. Her research to date has focused on the geographies of citizenship, education, behavioural forms of governance, and the influence of applied and popular neuroscience on policy and practice. She teaches on the spatial politics of welfare, work and wealth.

Fiona M. Smith is a Lecturer in Human Geography in the School of Social Sciences, University of Dundee, UK. Her current research focuses on themes of gender and youth geographies, relational practices of care, volunteering and resilience, and health and wellbeing in relation to greenspace. She is co-author of *Geographies of new femininities* (with Nina Laurie, Claire Dwyer and Sarah Holloway).

Lorraine van Blerk is Professor of Human Geography at the University of Dundee, UK. As a social geographer working mainly in sub-Saharan Africa, Lorraine's research explores questions related to understanding how the practices and conditions of growing up influence young people's lives. Her work draws on feminist theory and participatory philosophy and is rooted in a desire for justice and equality in addressing issues of poverty. Lorraine has written over 70 publications, many directly concerned with children and youth. She is also co-author, with Kathrin Hörschelmann, of *Children, youth and the city* and co-editor with Mike Kesby of *Doing children's geographies*.

Kayleigh van Oorschot, MSc, works in the Netherlands as a project manager in the educational sector with a focus on innovation. For the Netherlands School of Public Administration (NSOB) she has researched governance in the judiciary system, libertarian paternalism and the international application of 'nudge'.

Mark van Twist is Professor of Public Administration at Erasmus University Rotterdam, the Netherlands, and dean and member of the board of the Netherlands School of Public Administration (NSOB). He has published numerous contributions to books and journals, including in *Policy Sciences, Public Administration, Public Integrity* and *Public Management Review*.

Mark Whitehead is Professor of Human Geography at Aberystwyth University, UK. His work focuses on various aspects of environmental governance. His most recent research has considered the emerging impacts of the behavioural sciences on the design and implementation of public policy.

Acknowledgements

The editors would like to thank all the authors for their rich and varied contributions and for sticking with the book's development over some time. Our thanks are due to the organisers of the Fourth and Fifth International and Interdisciplinary Conferences on Emotional Geographies, held in Groningen in 2013 and Edinburgh in 2015. Papers given on emotions and governance at sessions we organised in Groningen provided the genesis of the book, and we were able to further develop themes in sessions in Edinburgh as the writing stage neared completion. Thanks to Katy Crossan at Ashgate for originally commissioning the book, and to Faye Leerink and others at Routledge for picking up the reins once we had delivered the manuscript. Particular thanks to Fintan Power for careful copy-editing. We are also very grateful to Val Britton for permission to use her artwork for the cover.

Ellie would like to thank Caitlin, whose imminent arrival stopped her coming to Groningen, but who has made life more fun ever since. Jessica would like to thank Eva and Manon for improving her emotional state. Fiona would like to thank her parents for their support in the final stage of the book's production.

1 Introduction

Governing with feeling

Jessica Pykett, Eleanor Jupp and
Fiona M. Smith

A 'more human' government?

In May 2015, following a return to power of a right-wing Conservative UK government, and resultant despair from the UK left about the failures of its political imagination, a new arrival appeared on the bookstands of popular political non-fiction. Entitled *More human: designing a world where people come first* (Hilton et al., 2015), the book is written by a former adviser to the current British Prime Minister David Cameron, and is a call to both the left and the right to radically rethink government, politics and policy. The book opens with an anecdote about the treatment of a family on a budget airline flight that frames the rhetorical question: 'Why can't we just treat each other with kindness and decency, like human beings?' (p. 2). What follows is an exploration of what this might mean across strands of government, business, education, urban planning, the food industry and more, with key themes of the need for more 'empathy' and care, smaller-scale and thus closer government, and less 'bureaucracy'.

From a left-wing perspective, the book is easy to ridicule. Nick Cohen (2015), for example, provided a rapid dismissal of Hilton's platitudes in light of the emotional misery created by Cameron's inhumane welfare reforms. It remains to be seen whether Hilton, architect of the previous government's widely dismissed 'Big Society' approach (Runswick-Cole and Goodley, 2011), has produced an intervention which will have much traction within contemporary political and policy debates. Notwithstanding this, we would argue that his book encapsulates and expresses something that has been apparent for at least the last decade of political discourse in the UK and elsewhere. It reflects a new enthusiasm for an emotionally attuned approach to government which sees emotions as constitutive of the very workings of government and policy. *More human* therefore forms part of a broader genre of writing that has popularised emerging ideas about the emotions and the self within government and governance. Like many of its predecessors, it calls up particular advances from neuroscience, behavioural economics and social psychology to rethink what we know and understand – often evoked via science and 'hard' evidence – about human nature. It then reimagines the more 'humanised' policies which should follow, whether that is rerouting traffic in New York City or implementing more home visiting schemes in healthcare. Indeed, Hilton draws attention to the international nature of this phenomenon, even whilst remaining primarily focused on UK policy and politics.

Putting aside the politics of the arguments, which will be explored further below, an interesting starting point is to reflect on the inclusive and appealing nature of Hilton's propositions. On one level, a call to be 'more human' is hard to disagree with (although, as Terry Eagleton (2015) witheringly points out, 'rape and genocide are human too'). There is something inherently engaging about considering emotions, the intersubjective warmth and friction of human inter-actions and feelings, in shaping public policies aimed at increasing human happiness. It is easy to understand why this kind of policy ideology has been popularised – arguably worldwide – above other strands of economic and political analysis (which nonetheless may remain important within policy regimes). Talk about feelings and human nature draws us in. Indeed this strand of governance and management is attractive in a commercial sense too, with international institutions and organisations willing to pay considerable sums for toolkits and training in new forms of emotional literacy.

While this book is primarily concerned with the 'emotional state' of the UK, it is clear that emotional governance and emotional politics have global resonances which remain under-researched. We may take examples of the emotionalised rhetoric in US political debates as indicative of the growing salience of public and political expressions of emotion, for instance Donald Trump's recent invocations of fear, hate and disgust of both migrants and America's Muslim population as part of his 2016 campaign for the Republican presidential nomination. Or we might consider the significance of moving images of President Obama's tearful responses to ongoing gun crimes against innocent children, or the anger inspired by the seemingly intransigent racial injustices faced by Black US citizens. Further examples of such emotional politics might be found in the recent global appeal of wellbeing and happiness measures across several nation states, posed as a corrective to the dominant focus on economic growth at the expense of 'more human' considerations. So, too, emotionalised work and relationships might be viewed as essential to sustaining social movements, whether resistive or not (Aminzade and McAdam, 2002), and to geopolitical issues of conflict, security and peace-building (Crawford, 2000). In the summer of 2015 the circulation and affective force of an image of a drowned child in the Mediterranean within the Western media seem to have generated an unexpected shift in international policy as well as popular attitudes about migration, although these feelings may in themselves obscure the more complex politics of migration, international discord on the provision of refuge and national immigration policies themselves shaped by highly emotional dynamics (which is explored by Forkert et al., Chapter 12, in this volume).

This book explores these issues via focusing primarily on one context, the UK (but with other non-UK comparisons). However, it is also important to note that within the UK state there are diverse geographies of policy-making and practice, whether that is in the devolved administrations of Scotland, Wales and Northern Ireland, meaning that a range of policies apply specifically in

England rather than the whole of the UK, or whether it is reflected across a range of local administrations in relation to local government. Thus a number of the chapters engage with emotions in policy, policy-making and citizenships in the devolved administrations of Scotland (Anderson, Chapter 6, Smith et al., Chapter 15) and Wales (Pykett et al., Chapter 5), while some examine aspects of local governance (Collins, Chapter 13).

Overall, the book seeks to unpack the seductive attraction of feelings-talk itself as something we also need to acknowledge, lest we describe or evoke emotions as an end in themselves (Pile, 2010). However, we seek to go beyond such allure to explore the economic and political values and currencies of emotions. We ask what happens when emotions become entangled with government and policy. We are interested in advancing understandings of how emotions are put to work, how they circulate and perform, whether in the service of government agencies, policy practitioners or citizens and communities. We consider how emotions create and sustain new relationships of power and modes of governing, but also how they might disrupt governance regimes, surface in unexpected ways and places, generate new identifications and solidarities, and perhaps shift policy and politics, as in the example of refugee policy above.

The book therefore has at its centre ambivalence, contradiction and the always unstable nature of emotions and their geographies. It unpacks the cultivation and reproduction of embodied and emotional notions of selfhood experienced by the citizen, policy-maker and public service worker, arguing that *emotionalised states* are more than a new set of vocabularies, techniques, objects or sites of governing. Rather, they signify contradictory political imperatives. On the one hand, there is a common-sense, evidence-based and apparently progressive drive to govern '*more* humanly', in sometimes highly personal and interventionist ways. On the other hand, emotionalised states and their governments also somehow govern *less* – by rejecting the modernist social contract between state and citizen, renouncing the notion of a highly rationalised state bureaucracy in favour of personalised forms of self-government. This is even more evident given that this novel enthusiasm for emotionally attuned government comes at a time of radical state roll-back in neoliberal democracies such as the UK and USA.

However, as already noted, within and in between these explicitly emotionalised techniques of governance, other kinds of emotions inevitably circulate and may contest their rationalities. Therefore, as well as the work involved in *governing the emotional states* of citizens, we are interested in *emotional forms of governance*, highlighting the ways in which the work of state agencies, civil servants and public services always involves emotional negation, excess, dilemma, rhetorical fantasy, as well as emotional celebration and commitment. In charting the politics of both these emotional forms of governance and related efforts at governing emotional states, the book identifies crucial points of contestation and activities of resistance which explicitly address the aforementioned seductive allure of 'more human' modes of governance.

The book therefore stands at the intersection of, on the one hand, a burgeoning social science literature that places emotions centre stage across a range of fields of enquiry and, on the other, a simultaneous interest in emotions from politicians and policy-makers. Its contributors introduce a range of empirical sites, from school inspection regimes to immigration policy, early intervention with children and urban regeneration initiatives. In order to explore emotions within governance we examine spaces of *public policy-making* (Part II), *public service delivery* (Part III) and *citizenship and participation* (Part IV). The book is organised along these lines, after two chapters in Part I that explore broader issues around researching emotions and governance. Before introducing these, we turn to consider some theoretical and methodological trajectories around social science research on emotions.

Emotions within social science research

Although this book does not aim to provide a philosophical or sociological summary of existing research on emotions, it is helpful to provide some background context on the major developments in this area in order to avoid overstating the novelty of our focus on emotions within governance regimes. Historical and sociological accounts provide ample commentary on the sixteenth-century Reformationist 'disenchantment' of emotional life and its later reflection within the emerging 'spirit' of capitalism (Weber, 1974, cited in Williams, 2001: 20). So, too, classic texts recount the civilising imperatives associated with emotional management and manners developed throughout the modern Enlightenment period, during which emotions as bodily sensations were viewed with mistrust and were also highly classed (Elias, 2000 [1939]). Sociologist Simon Williams provides a particularly helpful journey through the place of the emotions in twentieth-century sociological thought, illuminating its vitalism, sensuality/sensibility and its focus on irrationality and the passions through the founding work of Weber, Durkheim, Simmel and Marx. He notes a particular renewal of interest in the sociology of emotions during the 1980s and 1990s, stemming from the contributions of feminist and queer theorists. They established more embodied and gendered accounts of emotions, as well as incorporating neuroscientific advances in understanding the centrality of emotions to processes of cognition. Such advances in social *and* scientific theory provided a radical challenge to Western binaries of (feminised) emotion and (masculinised) reason, which paved the way for the development of a sociology which ascribed value to both the material and the discursive aspects of personal emotion. Hence the work of sociologists including Deborah Lupton and Bryan Turner has been important in promoting a view of emotions as 'embodied sociality' (after Lyon and Barbalet, 1994: 48, cited in Lupton, 1998: 4). This perspective aims to confront the limitations of 'overly' socially constructionist accounts of emotions (such as that offered by Rom Harré) and structuralist perspectives (e.g. Marx, Durkheim and, later, Hochschild). Such approaches are said to give little time to the physiological or embodied aspects of emotions, and to the potential for individual actors to shape their own emotions outside of predefined rules of comportment (see Lupton, 1998: 21). But at the same time this notion of 'embodied sociality' put the social contexts of emotional experience centre stage, appreciating the contingency of embodied emotions across cultures and over time, acknowledging the importance of norms in shaping the experience of emotions and recognising the socially situated (gendered, racialised, classed) nature of the interpretation and meaning of emotions.

The embodied nature of emotions was to become one of the most salient features of the sociology of emotions at the end of the twenty-first century, and reflected the aspirations of sociology itself to become, in some sense, 'more human'. For example, Elizabeth Grosz's emphasis on emotional flows and 'leaky bodies' pointed towards the fleshy materialism of the human body as an animal (Grosz, 1994, in Lupton, 1998: 87), troubling boundaries between self and other, inside and outside, order and chaos. Feminist theorists spoke out vociferously against the apparently unembodied rational man and feminism's own dismissal of biologism and neurologism (Wilson, 1998; Sedgwick and Frank, 1995). They sought to reclaim a definitive materiality of emotions through an attention to its biopyshical and somatic dimensions and through the development of theories of affect. This movement was further buoyed by insights from the emotion sciences, affective neurosciences and psychology. Work as diverse as that of neuroscientist Antonio Damasio, who provided a new account of the brain functions of the emotion of fear, and psychologist Silvan Tomkins, who set out to identify and classify inherent biological affects, has been influential in rethinking the potential value of the biological sciences to understanding the sociology of emotions. This move has also been subject to critical scrutiny. For example, Papoulias and Callard (2010) provide an account of the problematics posed by the adoption of biological concepts in cultural studies. Furthermore, there has by now been substantial debate as to the precise distinctions between affects and emotions (see Pile, 2010). Nonetheless, it is clear that the 'affective turn' has had a significant impact on the way in which sociologists, cultural theorists, contemporary neuro-philosophers, human geographers and political scientists alike conceptualise embodied emotions. Political scientists such as Martha Nussbaum (2001, 2013) have insisted on the centrality of emotions to ethical reasoning, human intelligence and political culture, whilst others argue that the neurosciences herald a new era of political theory which should more closely examine the neurobiological and genetic precepts of social and political behaviour (Hibbing, 2013) and ultimately protect 'policy from the undulating passions of the madding crowd' (Neuman et al., 2007: 7). Nikolas Rose (2013: 3) has proclaimed that 'the biological century' in which we now live requires a new kind of sociology and a new politics of life.

And yet these biologically driven accounts of the passionate body politic and the neurobiological political subject have also been widely resisted. For many, there is still a dangerous political scientism at the heart of accounts which seem to reduce complex social and relational feelings to their biological correlates, as if they were unmediated by social structures, history, culture or intersubjective experience. Thus for social psychologists, including Margaret Wetherell (2012), whose approach informs that of some of the contributors to this book (Clarke,

Chapter 9, Newman, Chapter 2, Jupp, Chapter 10 in this volume), there is a focus on human emotion as a set of *affective practices*, which contributes to, but can also disrupt, wider *affective patterns*. This highlights the distributed nature of emotions which exists between individual subjectivities and shapes the structures and institutions of collective life (see Hunter, Chapter 11 in this volume).

An important issue for such analysis is thus around *how emotions are understood to move* beyond individual subjectivities. These are questions raised by Ahmed (2004), who proposes an understanding of 'affective economies' to describe the social and public practices of emotions as 'sticky', moving between signs, bodies and objects, 'surfacing' in unexpected ways or recalling past experiences that may have seemed forgotten. The idea of affective contagion (Blackman, 2008) offers a complementary way to conceptualise the movement of emotions and feelings, focusing on the very nature of the sociality of emotions in action. Blackman argues that recent materialist forms of cultural theory have failed to get beyond questions of social influence (inside/outside; structure/agency), a core problematic in social scientific research. Furthermore, she shows how the notion of 'suggestibility' has been reimagined through contemporary social theorists of emotions (listing Latour, Thrift, Massumi, Brennan, Connolly) as an inherently weak and inferior mental state. The danger here is that this can be used in describing how people and situations can be manipulated, thus inadvertently reviving an atomised ideal of the individual subject who should be free from social influence. What Blackman's work importantly reminds us is that any social science conceptualisation of emotional practices also requires a clearly articulated theory of power and governance. Only in doing so will it be able to avoid replicating hierarchical notions of human subjectivity – in this case, along different degrees of suggestibility.

Therefore, in taking the regulation, modification, mobilisation, management and governance of emotions as the core issues, *Emotional States* is sensitive to the production of hierarchical foundations of emotional skills and competencies. Rather than focus on questions of the causation, basis and definition of emotions, this book foregrounds a re-contextualised approach to human subjectivity. We argue that attention needs to be paid to how social, institutional and strategic practices are rendered meaningful, producing particular kinds of emotional subject positions which are culturally, historically, politically and, importantly, geographically specific. In the next section we expand on the particular contribution of the book to debates on emotions, policy and politics, highlighting theoretical and conceptual resources from human geography and feminist theory, as well as expanding on the theme of contradiction and ambivalence as a key aspect of our analysis.

Approaching emotional states: feminist geographies of governing with feeling

To this point we have seen how different conceptions of emotions are a matter of considerable debate across the social sciences. Our concern in this book is not to define what emotions are, but to describe and analyse what they do and what is done to them in specific policy and governance contexts. Emotions are powerful, in more ways than one, and it seems reasonable that different strands of research would have different orientations towards this power. As already indicated, in this collection we hold on to both a critical analysis of emotions as aspects of certain forms of governance and politics, and also an analysis of the ways in which emotional registers of hope, possibility, resistance and solidarity can 'surface' in unexpected places, leading to possibilities of transformation and change. In particular we wish to make a case for working across social policy and geography disciplines to understand the differentiated emotional dynamics of governance processes in spaces and places.

Within human geography, Davidson and colleagues (2005) reported on the emergence of an 'emotional turn' a decade ago. The feelings and emotions that circulate around spaces and places clearly matter within governance and policy regimes at various scales: from the 'affective architecture' of new kinds of educational setting (Kraftl, 2006), to broader propositions around 'localism' that seek to work with residents' attachments, feelings of belonging and knowledge of particular places (Jupp, 2013). At the same time, emotions are also part of the relationships between distant places, and can hold together the geographies of 'imagined communities', institutions and collectivities, whether that be via the affective economies of online activism or in determining who qualifies as a citizen. As pointed out elsewhere (Jupp, 2013), the language of critical social policy analysis of emotions, which may consider aspects of being inside/outside, included/excluded and intimacy/detachment, points to the symbolic but also (economically) material geographies of power and practice in processes of governance. In approaching the slippery terrain of emotions, state, policy and governance, we wish to demonstrate how using geographical perspectives can enable moves between detailed empirical analysis of sites and spaces, consideration of the mobilities, relational dynamics and 'fixing' of emotional landscapes, and a wider critical reading of the politics of the current moment of emotional governance. To return to Wetherell's terms above, affective 'practices' and 'patterns' can also be thought about in terms of the more intimate and the wider emotional geographies that might be discerned in approaching governance regimes.

Researching these different kinds of emotional geographies clearly has implications for methodologies. Linked to the growing subfield of emotional geographies (Bondi, 2005), developments in research methodology are particularly pertinent to the analysis presented by contributors to the book, and each chapter reflects on the methodological demands and innovations required to engage in research on emotional governance. In different ways, the chapters pay close attention to the emotional language, encounters and performances embodied at each site of emotional governance, including elaborating on the place of the researchers themselves in mediating, interpreting and representing affective practices and patterns. As well as interviews and focus groups, therefore, methods include ethnography at various scales (Anderson, Chapter 6, Lea et al., Chapter 7, Dobson, Chapter 8), critical analysis of policy and governance, informed by psychoanalytical approaches (Clarke, Chapter 9, Hunter, Chapter 11), autobiographical reflection (Jupp, Chapter 10), participatory methods (Pykett et al. Chapter 5) and the fictionalised staging of an embodied emotional subject (Griffiths, Chapter 14).

Geographical perspectives and methods, therefore, can enable a multi-scalar approach to emotions within governance regimes that can move between the embodied negotiations of welfare encounters and the wider affective economies of media and policy texts (see in particular Jupp, Chapter 10 in this volume). This seems important given the complexity and ambiguity of evaluating the politics of emotions in these contexts, a politics which crosses the research process itself within the formulation of policy strategy, in sites of public service delivery and in a range of citizen-engagements with the state. Indeed although the chapters of this book are divided into sections reflecting these divisions, we aim to show their interconnections and instabilities.

One such set of instabilities stems from the links between forms of feminist theory, activism and research commitment, 'making the personal political', on the one hand, and the adoption by governments of registers of governing and delivering services which seem to chime with these forms of theory and activism. This problematic might be seen as part of the wider problematic of the relationship between feminism and neoliberal governance (see Fraser, 2009). For example, a set of discussions around the 'relational state' (IPPR, 2012) has argued for the recognition and promotion of individual relationships and the emotional labour involved in the delivery of public services. Promising a break from understandings of the state as 'transactional' or 'top-down' this discourse is also linked to the promotion of 'co-production' in public services and the invocation of qualities of 'care', 'compassion' and 'love' (Aitkenhead, 2013) in public sector work.

Several of the contributors to this book (see Newman, Chapter 2, Enright et al., Chapter 3 in this volume) therefore investigate the potential 'co-option' of the feminist pursuit of gender equality and justice by a policy agenda which promotes certain forms of sociality, full of emotional sentiment but perhaps stripped of political radicalism. Indeed we would argue that the emergence of emotional registers in policy discourses may actually signal the potential 'emptying out' of political discourse away from issues of justice, material equality and political recognition. Whilst notions such as 'the relational state' (IPPR, 2012) and 'co-production' (Needham and Glasby, 2014) in public services may seem to stem from an ethics of care and an emphasis on intersubjective negotiation, the 'relationality' of the 'relational state' and the actors involved in 'co-production' can be presented as strangely undifferentiated. Indeed the imperative to govern in ways that are 'more human' seems strangely non-descript, almost entirely ignoring emotions which might be conveyed as negative. Despite all this, both Newman and Enright and colleagues (this volume) suggest possibilities for new kinds of creative labour and professional development, for women in particular, within this landscape of relationality and co-production, and Enright and colleagues in particular caution against overly sweeping critical readings of such a policy climate (see also Smith et al., Chapter 15, on these debates around the use of voluntary labour in service delivery).

Nonetheless, emotional resources and repertoires are of course in themselves linked to wider inequalities: who is able to muster and display appropriate emotions within encounters with the state, and whose performances are seen as 'abject' (Tyler, 2013)? We must consider how emotional forms of governance play out in unequal contexts, as well as how they might reproduce specific forms of inequality. The emotional resources, skills and 'care-full' commitments required by co-produced public services are not necessarily shared between citizens equally. Similarly, the psycho-economic impetus of the 'behaviour change agenda', as one form of emotional governance, tends to divide citizens into those capable of rational thoughts and social groups deemed more vulnerable to impulsive and irrational behaviours (Jones et al., 2013). Many of the chapters in this volume are UK focused, but, as is made clear by the contribution on Netherlands housing policy by van Oorschot and colleagues (Chapter 4), emotional governance has specific resonances in different nation states – pointing to the need for further comparative research which investigates differences across national polities, cultures and economies (as discussed by Clarke, Chapter 9, in this volume), at the same time as recognising the role of policy transfer and internationalised agendas in shaping contemporary emotional states.

This book therefore considers the ambivalent range of emotions that characterise public policy-making, implementation and participation at the current time. Indeed, in evaluating emotional forms of governance, it is imperative to set the contemporary UK policy interest in personal relationships in the context of the current cuts to and fragmentation of public services, where the whole notion of the public sector seems in doubt. Is it possible to form sustained relationships for policy delivery without a stable welfare system? And does the focus on emotional rather than other kinds of resources (i.e. financial, material) serve to further undermine such a system? There is an important role therefore for critical social science in pointing out the wider, and unequal, social, economic and political contexts which frame emotional and affective encounters between specific citizens and states. Otherwise there is a danger of thinking that, as Eagleton (2015) ironically writes, 'the role of the social sciences is to find out what makes us tick so that we may become more satisfied producers and consumers'.

Structure of the book

Part I: Approaching emotional governance: feminism and gendered labour

The first section of this book consists of two chapters which examine the overall politics and potentialities of new forms of emotionally inflected governance and professional practice, focusing on women's gendered labour. Contradiction and ambivalence are key themes in both chapters. In Chapter 2 Janet Newman provides an important steer towards the potential significance of emotional governance, as well as the dangers of overemphasising it as some kind of straightforward and fundamental shift, given that rational-technical models of policy are still dominant in many fields. The chapter positions neoliberalism as a paradoxical phenomenon and describes the way in which emotions are incorporated into a neoliberal ideology (and wider political landscape). The chapter identifies within neoliberalism a certain 'rage' against welfare, bureaucracy and irresponsible, 'immoral' ('inactive') citizenship. At the same time, neoliberalism builds on and supports emotional values (e.g. community, morality, care and solidarity) associated with the feminist movement. This presents a certain paradox, which the chapter seeks to move beyond by focusing on forms of practice (here among women working in the voluntary and public sector) as a complex and often painful form of 'affective performance' (after Wetherell, 2012: 226), 'unevenly patch[ing] together different identifications and allegiances' (Newman, Chapter 2).

The chapter by Bryony Enright, Keri Facer and Wendy Larner (Chapter 3) examines some of these same tensions within research and academic labour itself, turning the spotlight on imperatives for 'co-production' within the academy itself, taking the example of female academics working on community-based research projects funded under a UK Research Council grant scheme. The chapter discusses the contradictory ways in which the emotional landscapes of the academy are being reshaped by imperatives for co-production with marginalised research subjects. Despite some of the difficulties and limitations of this, the chapter is clear that 'critiques of co-production as the marketisation and instrumentalisation of the academy are clearly insufficient to explain the motives, practices and consequences of the relational academic labour . . . achieved' (Enright et al., this volume). Instead the research programme emerges as a space of relative freedom, creativity and emotional transformation for the women academics involved, as well as potentially their co-researchers beyond academia.

Part II: Emotions in public policy-making

As already noted, emotions have come to be known not just as means to specific political ends but as strategic objects of governance in and of themselves. For example, the explicit measurement of national happiness has engendered fervent warnings against the reductive economisation of emotions and wellbeing (Davies, 2015; Frawley, 2015). To a degree this reflects earlier concerns of an emerging 'emotivist ethic' in both public life and state culture (Nolan, 1998: 5). Such an ethic places emotions at the foreground of our self-understandings and moral terms of reference, which, Nolan argues, lapses into state–citizen relationships and encounters based on narcissism, self-help and personality obsession.

Meanwhile, several national governments, notably the US, the Netherlands, Denmark, Australia and the UK, have invested in and experimented with new forms of behavioural governance (Jones et al., 2013). These 'nudge' type policies are based on the targeting of the emotional drivers of decision-making, and promote governing through directing affects and/or shaping the choice environments in which people act (van Oorschot et al., Chapter 4, Pykett et al., Chapter 5, in this volume). Such techniques explicitly seek to govern affective, non-rational behaviours, raising questions about the role of the state in *governing emotional states* and addressing citizens as complex, often conflicted humans with feelings. These behavioural governance techniques rely on specific scientific insights on human behaviour from disciplines including behavioural economics, psychology and neuroscience, and are primarily concerned with the apparently predictable psycho-physiological determinants of emotionally driven action rather than emotional meaning-making or discursive enactment of emotions.

The chapter by Kayleigh van Oorschot, Menno Fenger and Mark van Twist (Chapter 4) provides an overview of 'nudge' type policies and 'choice architecture' (Thaler and Sunstein, 2008) and examines the potential for such approaches to reinvigorate the failing Dutch housing market by structuring choices within the always-already-emotional geographies of home and home buying. It points to practical and ethical dilemmas in applying such approaches, which despite the term 'nudge' involve quite significant forms of economic and social intervention. In Chapter 5, Jessica Pykett, Rachel Howell, Rachel Lilley, Rhys Jones and Mark Whitehead develop this terrain through a discussion of a participatory form of research intervention that involved civil servants and policy-makers engaging with 'mindfulness practice as a means for participants to engage more critically with insights about the nature of human behaviour and how it can be changed'. This project therefore sought to highlight some of the critical and ethical dilemmas around 'nudge' techniques for policy-makers themselves and foregrounds some of the political tensions inherent in researchers' engagement with this terrain, developing a mode of 'positive critique' involving practical engagement rather than critical deconstruction. Rosie Anderson's chapter (Chapter 6) also examines how policy-makers and civil servants understand emotions and subjectivities in relation to the targets of their policies and how they experience the affective spaces of encounter with them. The chapter emphasises the value of taking a practice-based approach to governance studies, thus resonating with Newman's chapter above. Anderson discusses instances of the antipathy civil servants seemed to feel towards emotional displays, but at the same time shows how the policy-making process is fraught with emotion. Interviews and ethnography with civil servants show complex relationships between emotions, 'reality', legitimacy and evidence, demonstrating some of the 'affective performances' (Wetherell, 2012) at stake and disrupting 'assumptions about the linear and purposive nature of policy work'.

Part III: Emotions in public services

Moving from the formulation of policy strategies to the related but more specific techniques of policy intervention, as noted above, a set of discussions around the 'relational state' (IPPR, 2012), co-production and personalisation in public services has variously promised to reshape relations between citizens and public sector professionals within more emotionally attuned and personally responsive ways. As also already intimated, in a moment of public sector cuts, this can be seen critically as low-cost policy reform – dependent not on professionalisation, funding or process change, but specifically on rehumanising state–citizen relations, bringing empathy 'back in' and emphasising the emotional skills of those engaged in public sector work (Needham and Mangan, 2014; Anderson, Chapter 6 in this volume; Dobson, Chapter 8 in this volume). And indeed, as previously suggested, these 'more human' qualities often invoke a benign and de-politicised version of human relations.

In fact, of course, this relational state exists concurrently with a state which both invokes, produces and shapes a far more ambivalent and troubling range of emotions than those of empathy, compassion and care. In this part of the book we examine a range of sites of public policy in which emotions including paranoia, anxiety, anger, shame and apathy circulate. At stake within them are often explicit governance attempts to shape forms of selfhood, what we have called 'governing the emotional states of citizens'. Within these sites we see that certain emotional competences, relations and resources are valued above others, reflecting in part some of the new agendas around behaviour change discussed in the previous section, as well as longer-standing attempts by the state to shape subjectivities.

The chapter by Jennifer Lea, Louise Holt and Sophie Bowlby (Chapter 7) demonstrates the troubling and exclusionary nature of the operation of such policy initiatives, using the example of school-based practices and curriculum offerings aimed at social and emotional development. Their research shows how this can be significantly exclusionary for children designated as having behavioural, emotional and social difficulties, exclusions which play out within the specific geographies of the classroom. They argue that such practices suggest 'new limits to citizenship . . . based on the ability to develop a relation of self-governance of the emotions'. Rachael Dobson (Chapter 8) also demonstrates in her chapter the complexities of attempts to shape subjectivities via social policy through a study of homelessness practitioners, investigating how shifts in housing/homelessness policy, regulation and governance are creating new dilemmas for front-line staff. The changing regulatory framework relies on staff making particular assumptions about homeless people's lives, specifically their assumed goal of 'independent living' which is regarded as something of a panacea for a complex array of problems. The chapter also shows how new kinds of relationships between staff and homeless people are being forged, and hints at the behavioural interventions which are now *de rigueur* in homelessness service provision, especially an emphasis on 'transitional and transformative visions' in approaching vulnerable people and their futures.

The impacts of targets allied to attempts at transformation of various kinds are also at stake in John Clarke's contribution (Chapter 9), which demonstrates the value of a psychoanalytic approach to analysing the regulatory regime of school inspections in the UK. It develops an argument around the notion of 'emotional excess' and the circuit of emotional dynamics which mediate inspection judgements in a specific political context. These emotions include anxiety and paranoia, affecting not just the schools being inspected but the inspectors too, suggesting the 'anxious subject that forms in the shadow of the incitement to be responsible, independent and empowered'. Eleanor Jupp's chapter (Chapter 10) also focuses on the circulation of emotions and affect, in this case in relation to family policies, showing how they move beyond the spaces of encounter between public sector professionals and service users to encompass wider emotional geographies, including those within policy discourses and media representations. Although the feelings of professionals working with disadvantaged families are inevitably shaped by these wider emotional geographies, the chapter is also suggestive of the possibilities for different emotional registers to emerge from more concrete and compassionate accounts of the situations in which vulnerable families find themselves.

Part IV: Emotions of citizenship and participation

As should be clear from the chapters above, it is no longer (if it ever was) sufficient to consider emotions within governance regimes as occurring within the 'implementation' stages of rational, technical policy, as some paradigms for writing about encounters with the state and professionals might have once suggested (e.g. Lipsky, 1980). Instead, as argued above, emotions are increasingly explicitly understood to form part of governance, from the care and engagement needed for the co-production of services to the explicit shaping and harnessing of emotional dynamics within policy initiatives aimed at shaping conduct. In this section of the book, we examine the feelings associated with being a citizen encountering the emotional state in a number of ways – reacting to policy initiatives and change, being communicated with in emotional registers, and having certain emotions, for example civic pride (see Collins, Chapter 13 in this volume), evoked and shaped as an explicit part of local governance or proposals such as the UK government's 'Big Society' initiative which sought to draw on citizens' emotional engagements with and attachments to local place and community to incorporate forms of volunteering and citizen engagement into service delivery and governance.

Of course attempts by government to engage citizens in such ways, from more participatory democracy to volunteering schemes, are not new. Citizens' encounters with the state have always been laden with emotions. However, the current policy climate in the UK and elsewhere perhaps makes these encounters particularly complex and ambivalent (Smith et al., Chapter 15 in this volume), whereby questions over whose emotions are being evoked, valued and to what ends are increasingly politicised. The circulation, movement and representation of emotions are key issues in a number of chapters in this section, as well as their unstable and unexpected nature.

Shona Hunter's chapter (Chapter 11) considers the emotional landscape of the English National Health Service (NHS) at a moment of crisis, not in terms of how it feels to work within it, but in terms of how it is perceived and experienced as what she terms 'an affective formation' by politicians and citizens mobilising to 'save' it. She uses the notion of fantasy as a means to understand the relational politics of the NHS, specifically how an oppressive British imperial history is obscured through the 'feeling work' that goes into reproducing the ideal of the NHS as a solidaristic and equal British institution. She argues the NHS functions as a vehicle of fantasy for users and campaigners and politicians alike, obscuring the complex relations of power that form the NHS, and thus actually rendering its 'defence' more difficult. Staying with the terrain of multiculturalism and questions of British identity, Kirsten Forkert, Emma Jackson and Hannah Jones (Chapter 12) focus on the misalignment between an intentional form of 'emotional governance' and its reception in practice. The chapter discusses a recent UK government intervention around migration called 'Operation Vaken' in which illegal migrants were targeted through advertising hoardings urging them to 'Go home or face arrest'. The emotional geographies surrounding this campaign are revealed to be uneven and racialised, with many others feeling fearful beyond the intended group, but also with feelings of anger towards and solidarity with migrants being invoked. They use Bridget Anderson's (2013) idea of 'communities of value' in relation to whose feelings around this issue are deemed to 'count' from a government perspective, although emotions are also shown to be unstable around this issue in ways that open up potentials for activism and solidarity.

Questions of space and place and who belongs are also central to Tom Collins' chapter (Chapter 13) on the circulation of 'civic pride'. The chapter investigates how civic pride is used to govern cities and citizens to behave in particular ways, drawing out how civic pride is both celebratory and proprietary, and bringing this into conversation with political debates about localism in the context of austerity and economic insecurity. Hence the chapter posits cities as a kind of battleground 'of and for emotions', with different groups of citizens performing and appropriating different versions of civic pride. Mark Griffiths' chapter (Chapter 14) also focuses on the politics of producing active citizens through forms of emotional governance. The chapter introduces a set of conceptual and methodological tools around affect, embodiment and the self and uses a semi-fictionalised writing style to evoke the emotional dynamics involved in becoming a volunteer on a government programme. In the final chapter of this section, Fiona M. Smith, Matej Blazek, Donna Marie Brown and Lorraine van Blerk (Chapter 15) discuss what might be construed as an example of neoliberalised emotional governance in a model which complements more formalised structures of 'intervention' with the emotional labour of volunteers (active citizens) in a project offering mentoring to young people 'at risk' of anti-social and criminal behaviour. In echoes of the discussion about the risks/opportunities of the co-option of emotional work into the wider practices of the emotional, neoliberal state above (Newman, Chapter 2 in this volume), the chapter explores how the ambivalent relational spaces of mentoring are actively negotiated by the young people, their families, the volunteer mentors and the wider social policy agencies involved. It identifies how such actors simultaneously value the emotional labour involved and insist on the need to address its limitations.

Afterword

Cultural and historical geographer Elizabeth Gagen provides the final reflective afterword for *Emotional States*. Her own research has long explored the spatial dynamics of psychologised and emotionalised forms of childhood subjectivity.

Here she offers further commentary on the ways in which the emotions have become a new form of currency within public life and policy-making. But through drawing on Foucauldian-inspired analyses and by her close fidelity to historical detail, Gagen counsels against overemphasising the novelty of emotional forms of governance. Instead, she provides her own substantive account of two influential psychological research programmes which have shaped the prevailing discourses of emotional management, impulse and self-control. She elucidates the ways in which emotional governance from late nineteenth century bodily practices of child development to mid-twentieth century behavioural insights on self-control and desire have signified historically contingent shifts in emphasis which value specific forms of emotional comportment and subjectivity. These forms of governance are shaped by the neuroscientifically, behaviourally and psychologically inspired techniques associated with emotional forms of governance described throughout the book, through which we are now urged to achieve idealised emotional identities.

In bringing the diverse sites and practices of emotional governance together with analyses of emotional relations, affective practices and their geographical patternings, the contributions to this book point both backward and forward towards the constitution of emotional capital as a new kind of value generated, utilised and negotiated by institutions, individuals, social groups and researchers alike, and towards the need for critical and reflective analysis of the complexities of emotional states.

References

Ahmed, S. (2004) 'Affective economies', *Social Text*, 79(22–2): 117–39.

Aitkenhead, D. (2013) 'Troubled Families head Louise Casey: what's missing is love', *The Guardian,* 29 November 2013, at: www.theguardian.com/society/2013/nov/29/troubled-families-louise-casey-whats-missing-love (accessed 3 October 2014).

Aminzade, R. and McAdam, D. (2002) 'Emotions and contentious politics', *Mobilization: An International Quarterly*, 7(2): 107–09.

Anderson, B. (2013) *Us and them? The dangerous politics of immigration control*. Oxford University Press, Oxford.

Blackman, L. (2008) 'Affect, relationality and the problem of personality', *Theory, Culture and Society*, 25(1): 23–48.

Bondi, L. (2005) 'Making connections and thinking through emotions: between geography and psychotherapy', *Transactions of the Institute of British Geographers*, 30(4): 433–48.

Cohen, N. (2015) '*More Human* by Steve Hilton review – blue-sky writing', *The Guardian*, 1 June 2015, at: www.theguardian.com/books/2015/jun/01/steve-hilton-more-human-review-david-cameron (accessed 10 June 2015).

Crawford, N.C. (2000) 'The passion of world politics: propositions on emotion and emotional relationships', *International Security*, 24(4): 116–56.

Davidson, J., Smith, M. and Bondi, L. (2005) *Emotional Geographies*. Ashgate, London.

Davies, W. (2015) *The happiness industry: how the government and big business sold us well-being*. Verso, London.

Eagleton, T. (2015) '*More Human* by Steve Hilton review – freemarketeering is now called putting people first', *The Guardian,* 5 June 2015, at: www.theguardian.com/books/2015/

jun/05/more-human-designing-a-world-where-people-come-first-steve-hilton-review (accessed 10 June 2015).

Elias, N. (2000 [1939]) *The civilizing process: the history of manners and state formation and civilization*. Translated by Edmund Jephcott. Blackwell, Oxford.

Fraser, N. (2009) 'Feminism, capitalism and the cunning of history', *New Left Review*, 56(2): 97–117.

Frawley, A. (2015) *Semiotics of happiness: rhetorical beginnings of a public problem*. Bloomsbury Academic, London.

Hibbing, J. (2013) 'Ten misconceptions concerning neurobiology and politics', *Perspectives on Politics*, 11(2): 475–89.

Hilton, S., with Bade, J. and Bale, S. (2015) *More human: designing a world where people come first*. Random House, London.

IPPR (Institute for Public Policy Research) (2012) *The relational state: how recognising the importance of human relationships could revolutionise the role of the state*. IPPR, London.

Jones, R., Pykett, J. and Whitehead, M. (2013) *Changing behaviours: on the rise of the psychological state*. Edward Elgar, Cheltenham.

Jupp, E. (2013) '"I feel more at home here than in my own community": approaching the emotional geographies of neighbourhood policy', *Critical Social Policy*, 33(3): 532–53.

Kraftl, P. (2006) 'Building an idea: the material construction of an ideal childhood', *Transactions of the Institute of British Geographers*, 31(4): 488–504.

Lipsky, M. (1980) *Street-level bureaucracy: dilemmas of the individual in public service*. Russell Sage Foundation, New York.

Lupton, D. (1998) *The emotional self*. Sage, London.

Needham, C. and Glasby, J. (2014) *Debates in personalisation*. Policy Press, Bristol.

Needham, C. and Mangan, C. (2014) *The twenty-first century public servant*, at: http://21st centurypublicservant.wordpress.com/ (accessed 20 May 2015).

Neuman, W.R., Marcus, G.E., Crigler, A.N. and MacKuen, M. (2007) 'Theorizing affect's effects', in Marcus, G.E., Neuman, W.R., MacKuen, M. and Crigler, A. (eds) *The affect effect: dynamics of political thinking and behaviour*. University of Chicago Press, London, pp. 1–20.

Nolan, J.L. (1998) *The therapeutic state: justifying government at century's end*. New York University Press, London.

Nussbaum, M.C. (2001) *Upheavals of thought: the intelligence of emotions*. Cambridge University Press, Cambridge.

Nussbaum, M.C. (2013) *Political emotions: why love matters for justice*. Harvard University Press, Cambridge, MA.

Papoulias, C. and Callard, F. (2010) 'Biology's gift: interrogating the turn to affect', *Body and Society*, 16(1): 29–56.

Pile, S. (2010) 'Affect and emotion in recent human geography', *Transactions of the Institute of British Geographers*, 35(1): 5–20.

Rose, N. (2013) 'The human sciences in a biological age', *Theory, Culture & Society*, 30(1): 3–34.

Runswick-Cole, K. and Goodley, D. (2011) 'The "Big Society": a dismodernist critique', *Disability and Society*, 26(7): 881–85.

Sedgwick, E.K. and Frank, A. (eds) (1995) *Shame and its sisters: a Silvan Tomkins reader*. Duke University Press, Durham, NC.

Thaler, R. and Sunstein, C. (2008) *Nudge: improving decisions about health, wealth and happiness*. Yale University Press, London.

Tyler, I. (2013) *Revolting subjects: social abjection and resistance in neoliberal Britain.* Zed Books, London.

Wetherell, M. (2012) *Affect and emotion: a new social science understanding.* Sage, London.

Williams, S. (2001) *Emotions and social theory.* Sage, London.

Wilson, E.A. (1998) *Neural geographies: feminism and the microstructure of cognition.* Routledge, London.

Part I

Approaching emotional governance: feminism and gendered labour

2 Rationality, responsibility and rage

The contested politics of emotion governance

Janet Newman

Introduction

Analyses of contemporary governance face a particular challenge. The current period of austerity and retrenchment in the UK and elsewhere means that the state appears to be shrinking, leaving citizens to assume responsibility for things it had previously provided. But at the same time, the state seems to be playing a more expansive role through its concern with the well-being and happiness of its citizens. In health, education, social care, neighbourhood work, in equality policies and policies concerned with environmental sustainability, human feelings and relationships are now at the centre of governing strategies.

This concurrent process of shrinking and expansion generates significant contradictions that are explored in this chapter. I begin by challenging the idea that governing with or through emotion represents a fundamental shift in governing logics. In *the politics of theory* I locate emotion governance in contested theorisations of neoliberalism, states and persons. But these 'big theory' questions cannot, I suggest, be understood without a simultaneous focus on the *politics of social practice*. The chapter draws on both registers to explore tensions in the concept of emotion governance itself. I show how emotion governance derives in part from social movements, and highlight the ambivalent politics that results. In *border work as social practice* I draw on my own research to show how new logics of governance were generated, in part, by women with a background in social movements. The dynamic tension between activist perspectives and governance logics challenges any sense that emotion management is an accomplished effect of new regimes of power. Rather it is actively negotiated in particular spaces and places.

Yet such negotiations have to be understood in the context of particular ideological climates. The chapter concludes by tracing how a *politics of rage* increasingly characterises austerity governance in the UK. This serves to sideline social movement perspectives, and to accentuate emotional registers of stigma, blame and shame.

The politics of theory: neoliberalism, states and persons

Contemporary social theory offers a series of rich metaphors intended to capture shifts in the technologies of governing (from government to governance; from social democratic to social investment; from welfare to workfare states, and so on). It is tempting to 'add' emotion to such forms of analysis, thus signalling a further shift from one mode of governance (in which personal and public are clearly distinguished) to another (in which states seek to reach out to the interior world of citizens). There are many emergent policy streams within the UK that seem to support the idea of such a shift: for example the emphases on personalisation and coproduction (Cahn, 2000; Hunter, 2007; Needham, 2011); on well-being or happiness (Dolan, 2014; Johns and Ormerod, 2007; Layard, 2005); on finding solutions to 'troubled' families or family breakdown (Davies, 2015); on preventing racial hatred and urban conflict (Jones, 2015), and many more. Such approaches draw on vocabularies of care, relationships, well-being, cohesion and happiness that suggest a radical departure from traditional welfare policies. But they are traversed by political ambiguities. They can be viewed as politically progressive, as a product of social movements – especially feminism – that foregrounded issues of personhood and what Williams terms a 'moral grammar of welfare "from below"' (Williams, 1999: 668). But they might also be understood as new forms of governmental power, congruent with the retrenchment of welfare states and new psychological orthodoxies. Underlying this paradox are contested conceptions of neoliberalism, of the state, and of the person that I briefly review here.

Neoliberal ideology is suffused with personal and social registers of governing that privilege – and seek to constitute – the self-governing subject, the responsible community, the developmental and entrepreneurial self (Brown, 2005; Brown and Baker, 2012; Rose, 1999). Neoliberalism, argues Rose (1999), is a discourse of community, of morality. Muehlebach (2012: 8) takes this further, arguing that neoliberalism requires, and seeks to constitute, compassionate, ethical and feeling citizens – what she terms 'moral neoliberals':

> The subject I am interested in performs two kinds of labors of care at once: it feels (cares *about*) and acts (cares *for*) at the same time. This subject is one that the state and many other social actors . . . imagine to be animated by affect rather than intellect, by the capacity to feel and act upon those feelings rather than rational deliberation and action.

The idea of moral neoliberalism, she argues, valorises feminised forms of work that the state no longer provides, thus mediating the effects of its own withdrawal. But the sentimental image of community, morality, care and solidarity also provide an anti-capitalist imagery that serves – paradoxically – to buttress neoliberal rule.

But how does this relate to actual processes of governing? Larner (2000) warns that neoliberalism must be understood as a 'hybrid political imaginary' rather than a unified and coherent political philosophy (p. 12), playing out differently in different contexts. It is versatile and malleable, actively appropriating projects and forces that appear be oppositional (a theme I return to later). The free market ideology of neoliberal economics is only loosely aligned to the neoconservative emphasis on family and community. The neoliberal project of constituting new forms of self-governing, responsible and perhaps moral citizens offers a range of discourses in which self and society, individual and community, are imagined and coupled in rather different ways. And governments pursue a range of political projects through diverse, and often incompatible, policy programmes. These different registers (neoliberal ideology, governmentality, policy) suggest different possible resolutions to the relationship between governance and emotion.

What of the state, the key actor in and instrument of governance programmes and policies? Dominant images of the state – especially in social policy and public administration – reference its institutions and structures, its concern with the relationships between means and ends, between policy 'levers' and desired effects, between governmental or bureaucratic incentives and policy outcomes. Any acknowledgement of emotion is squeezed into the 'implementation' stage of a seemingly logical, sequential policy cycle, where apathy, bloody-mindedness and other negative emotions (of staff or citizens) are assumed to get in the way of successful policy implementation.

Alternative conceptualisations of the state view it rather as an assemblage of discourses, practices, projects and technologies, all traversed by competing political forces and political projects (Newman, 2014; Newman and Clarke, 2009). The emphasis here is on a multiplicity of projects and policies that are not necessarily coherent or consistent. This idea of multiplicity offers one resolution to the paradoxical relationship between governance and emotion: some facets (such as economic management) may remain traditional and hierarchical, while others may turn to new strategies of governing – fostering relationships, engaging citizens in new projects of self-development. A rather different way of highlighting multiplicity is offered by Davina Cooper. The state, she suggests, has multiple identities and different facets: 'Bodies, work, purposes, powers, effects, responsibilities and form that combine, connect and become hierarchically ordered' (2014: 66). She also directs attention to the emotional life of the state itself: she writes of an 'emotionally contactful' state that is itself touched by wider events and that seeks to touch others, reaching out and engaging in 'attentive understanding' in order to embrace subjects (p. 62). The supposed rationality of state action, then, can be viewed as a well-rehearsed performance in which technical practices – action plans, implementing, monitoring – take place alongside other modes of contact that may involve encouraging, training, modelling, partnering and 'reaching out' to particular groups. And the state itself is suffused with emotional repertoires (disappointment, progress or frustration, rage and hatred) that undercut the veneer of purposeful action.

As well as contested conceptions of both neoliberalism and the state, unpacking the paradoxical relationship of governance and emotion requires an engagement with contested views of the person: as rational actor, as deliberative subject, as a bundle of behaviours, as an embodied 'feeling' subject. The process of governing has tended to privilege economic theory, resulting in dominant conceptions of welfare users and provider organisations as rational actors following market and other incentives (e.g. Le Grand, 2006). This offers a thin conception of human subjectivity – people are assumed to weigh up the cost and benefits of choice options and behave according to the incentives that policy offers. There is now a substantial literature that acknowledges flaws in the rational actor model, a literature that has helped inform the turn to behavioural economics and its focus on 'nudging' citizens towards making the right choices (Thaler and Sunstein, 2008). However, 'nudge' theory continues to offer a thin conception of the person, one who neither thinks, feels nor deliberates with others (John et al., 2013; Jones et al., 2013). Nudge strategies are assumed to work by bypassing the conscious mind to target environmental factors ('choice architecture') assumed to shape behaviour. Even where the desired effects of policy are saturated with emotion words – happiness, well-being, care, relationships – the means of 'delivery' remain instrumental, suffused with economic logics and behavioural change technologies. This focus on behaviour rather than feelings signals something of a perverse align-ment of governance and emotion. They belong to different discursive registers: as Tessa Jowell, then UK Minister leading the Sure Start programme, commented, emotion talk seems not to belong in the corridors of power: 'You don't talk about love in government' (Gerhardt et al., 2011). And they are not additive: it is impos-sible to add emotion on as an afterthought to rational-instrumental conceptions of how policy is formed and delivered, other than through an often token acknowledgement of 'emotional labour' at the 'front line'.

The analyses of neoliberalism, of the state and of the person I have traced here offer images of multiplicity and of ambiguity that undercut any idea of wholesale shifts in governance regime. Rather there is a need to explore the entanglements of emotions and governance through the specificities of social practice in specific sites and spaces of governing – the theme of the next section.

The politics of social practice

Recent years have seen some governance scholars turn away from 'grand theory' of shifting regimes towards a concern with how such regimes are interpreted and enacted. An 'interpretive turn' (Bevir and Rhodes, 2003, 2015; Hajer and Wagenaar, 2003) has informed a range of social policy research. This points to the potential displacement of existing policy discourses: social cohesion rather than social diversity and multiculturalism (Jones, 2015), active, responsible citizenship instead of welfare dependence (Newman and Tonkens, 2011), personalisation rather than universalism (Needham, 2011), consumerism and choice rather than solidarity and interdependence (Clarke et al., 2007), and so on. As well as tracing general discursive shifts, such studies also show how policy meanings are translated and negotiated.

However, as I have argued elsewhere, interpretive studies raise some troubling issues about the relationship between meaning-making, centred on the idea of the deliberative subject, and a post-positivist concern with questions of person-hood, identity and subjectivity (Newman, 2012b). Anthropological and ethno-graphic studies can offer a richer account of subjectivity and agency. For example, Catherine Kingfisher's study of 'welfare mothers' living in poverty in two contrasting sites (Aotearoa, New Zealand, and Alberta, Canada) shows how they engaged with processes of neoliberalisation. In both places poor single mothers were 'subjects of state interference designed to alter who they were as persons' (Kingfisher, 2013: 141). As such they were subject to discursive patterns based on stigmatising framings of welfare recipients. But these discursive patterns were subject to processes of translation by both welfare workers and the mothers themselves. Her research shows, for example, how the women in her studies negotiated contradictory welfare ideologies, navigating the tensions between neoliberal and neoconservative prescriptions, the former requiring that they adopt identities as independent worker–citizens, the latter focusing on their role as primary caregivers and dependent housewives. Other scholars have turned to psychosocial theory to explore the emotional and relational dynamics of policy formation and enactment (Froggett, 2002; Hoggett, 2000; Holloway, 2006; Hunter, 2015; Lewis, 2000). Yet others have drawn attention to the significance of affect and emotion (Wetherell, 2012) and to the importance of 'body work', showing how bodily practices and affective responses are inextricably entangled (Twigg, 2002; Twigg et al., 2011). Such work offers a more fine-grained analysis of how particular regimes of governance are enacted: how subjects respond to and engage with new governmentalities of personhood, and how individuals and groups generate new emotional or affective repertoires. It also brings to the fore some of the 'self work' that takes place as individuals seek to manage conflicting imperatives of governance or resolve the contradictions that are generated.

In an engagement with such questions Wetherell (2013) contrasts the work of Hochschild (1983) and of Anderson (2009). While Hochschild focuses on how actors regulate spontaneous and authentic emotional reactions ('the managed heart' of her title), Anderson's work centres on 'affective atmospheres': assemblages of technological, material and discursive practices. But while Hochschild highlights the *accomplishment* of emotional management, Anderson emphasises the *emergent* properties of such assemblages. This is in line with recent developments in geography and other social sciences, where assemblage is used to bracket essentialist conceptions of human agency and to challenge deterministic notions of change. But, Wetherell concludes:

> Neither 'feeling rules' nor 'affective atmospheres' are sufficient . . . to grasp the intertwining of emergent and accomplished affect. Flows of affect and emotion turn out to exceed simple characterisations as demonstrations of mostly active management or of mostly passive constitution . . . A way forward is to think about the relationship between unbidden affect and the active management of affect through the lens of social practice instead.
>
> (Wetherell, 2013: 222)

Programmes of emotion governance, from schemes to intervene with 'troubled families' or the coproduction of 'community well-being' to schemes to promote 'active citizenship', can be understood in such terms: as attempts at active management, the imposition of new 'feeling rules', that are cross-cut with what Wetherell terms 'unbidden affect'. But relationships between them can only be teased out through studies of social practice. In the next section I explore ways in which new 'feeling rules' are generated and negotiated, in the process highlighting the ambiguous role of social movements in the politics of emotion governance.

Border work as social practice

To engage with the politics of social practice I want to revisit a piece of my own research: a three-year study of the labour of women who had taken activist commitments into their working lives (with work encompassing informal, unpaid labour as well as paid employment). The initial study was based on interviews with 60 women across four generations, followed by a series of discussion groups and follow-up interviews. Participants were all based in England, but many had experience of political action in other nations. The research examined women working in community politics and campaign groups; in government and local government; in policy and the professions; in the voluntary sector and NGOs; in higher education, think tanks and research organisations. However, these categories are fluid; most participants in the study had fractured working lives that traversed different spheres of action. The research was completed in 2012 and reported in a series of publications (Newman 2012a, 2012b, 2013a, 2014). I have since extended the data through both follow-up and new interviews asking how activist women working within and beyond the state were negotiating the politics of austerity governance (Newman, 2013b).

The study begins to suggest how emotional registers of governing emerged out of activist, especially feminist, movements. Many of the participants had worked to bring more relational, person-centred and therapeutic registers into their engagements with policy and practice. They were, then, involved in *generating* and *embodying* new governing rationales, albeit in small scale, local or marginal spaces. Several had developed partnerships between government/local government bodies and 'communities'; their work was that of tutoring community representatives to 'speak to power', while fostering an emotional response among government actors in the interests of bringing about change. Others had promoted citizen involvement exercises that acknowledged, translated and mediated the affective and emotional responses of citizens (anger, dismay, hope) into language that policy actors might hear and pay attention to. Some had shaped new initiatives that were oriented towards a more relational and therapeutic style of intervention: Sure Start,[1] carer support groups (and later policies), self-help groups, well-being programmes and so on. Some spoke about their own management or leadership style as explicitly participative, relational and person centred. Many spoke the language of 'empowerment', for example through promoting forms of community mobilisation that aimed to reduce dependence on state funding and support.

Each intervention can be understood as challenging the personal/political boundary, opening up more of the self to governmental intervention (however benign). Each can be linked to the long reach of feminist politics, and its capacity to reshape governmental language and culture. Many became mainstream orthodoxies of governing, especially in the Blair years in the UK. But this does not imply that new emotional registers were an accomplished effect of policy. Tensions between rational-instrumental, means-ends policy styles and emotional registers were played out in evaluation programmes and funding regimes, in organisational hierarchies and in gendered labour processes. We might argue about how far later trends had rather different origins and were consistent with a more top down governmental approach (see for example Dolan, 2014, on the 'happiness' agenda). But the links between radical (often feminist) politics and governmental innovation continue: both norms of 'coproduction' and arguments for a more relational approach to welfare – even a relational state (Cook and Muir, 2012) – have their roots in left-leaning professionals, feminist inflected think tanks, activist organisations and campaigning groups.

However, such links between activist projects and changing forms of governing can invite a celebration of human agency, and offer a too optimistic image of what can be achieved in particular 'spaces of power'. By drawing on Wetherell's notion of practice, it is possible to suggest both the emotional and affective registers of policymaking and to highlight ambiguities in the enactment of new policy regimes. Practice based approaches, Wetherell argues, emphasise 'reflexive embodiment' rather than rule following:

> [P]eople adjust their affective conduct moment to moment, moving in and out of a sense of the self and the body as object, and as active subject . . . Affective performance, as part of working life, or indeed in any context, is likely to be more heterogenous, patched together, customised and interspersed with a wider range of embodied practices with no clear cut divide between the performances which mark 'public self' and those which distinguish 'time off'.
>
> (Wetherell, 2013: 226)

This perfectly captures the experience of the women participating in the research. Their practice was flexible and creative (Newman, 2013b). Yet, as Wetherell emphasises, this creativity was 'loosely determined' by what had gone before (2013: 234). As such participants in the research were both the object (of shifting governance regimes) and subject (crafting new developments and styles of policy and practice, informed by an activist, often feminist, sensibility).

The practice that I drew from the study is that of 'border work'. Most of those I interviewed were not in positions of formal power and did not have linear 'careers': new practices and projects tended to emerge in interstitial, informal spaces where activist or non-profit commitments encountered governmental experimentation and innovation. Their work cannot be understood as taking place 'after' policy, in the spaces of implementation or in the use of discretion by 'front line' workers; it was integral to the generation of new governing rationales. Affective performance thus offers a rather more expansive concept than that of emotional labour. It is based on a conception of the person as carrying and performing multiple identities; as at the same time an activist and paid worker, as an insider and an outsider, as a stranger and professional, as an individual, making her own decisions and compromises, and as part of collective entities. Such performances were patched together: in the language of some, this was expressed as stitching or knitting, crafting their work and their own, reflexive self. And in this patching, as Wetherell notes, there was little divide between public and private selves. The transcripts show how notions of work, politics and life, of care for others and care for the self, were weakly bounded, stitched together in particular ways at specific moments; see Gregg (2011) on the weak boundaries that enable work to enter into the 'intimate' spaces of home and personal life. Material conditions and personal responsibilities shaped the capacity to sustain such work. They often occupied marginal positions, engaged in forms of labour that rendered their work particularly precarious (Gill and Pratt, 2008).

One of the most interesting – and difficult – features of the study was that of hearing participants describe how developments they had worked for became mainstream. They often spoke of how things they had worked for over many years had been 'stolen', reinscribed with new meanings and inserted into neoliberal governing rationales with which they had little sympathy. This included their efforts to develop more human centred, relational and emotionally sensitive modes of engaging with citizens, users and communities. Reference was made earlier to Cooper's (2014) conception of the state as touching, reaching out, paying attentive understanding. This was the work of many of my participants. But rather than being implicated in the accomplishment of emotion management, by drawing on Wetherell I want to suggest a less deterministic approach – one that views emotion governing as a social practice generated by embodied subjects in ambivalent relationships with neoliberal power. The 'affective performance' of participants in the study required them to manage cycles of disappointment, despair and anger as well as periods of promise and hope. Many were involved in difficult negotiations with dominant ruling relations, struggling for change and then often watching that same change come to mean something different as it was taken up into mainstream policy – in the words of one, 'having the feminism stripped away'.

Austerity governance, populism and an affective politics of rage

Although I have emphasised the importance of the specificities of social practice, strategies of emotional governance cannot be divorced from wider political and ideological climates. The work of the participants in the study described above was situated in a changing political landscape. The welfare state was being challenged both by governments (in projects of state retrenchment) and by many professionals and activists (promoting the delegation of power to citizens and communities and more personal and relational styles of governing). This perverse alignment of activist struggles and austerity policies generated strain for those who had dedicated much of their lives to public/political work. The following extracts all refer to the period following the banking and financial crisis of 2008 and beyond as 'austerity' became the new orthodoxy:

We are all scrabbling for crumbs. I'm working with some other folk to set up a kind of social enterprise to keep the work alive, but it's a real struggle. We are trying to fill the gaps, to do some of what the state should be doing. But I don't really want to have to act like a business.

(advice worker, ex-voluntary sector, 2015)

I've always believed in challenging the power and authority of the state, and have tried to help disadvantaged groups take ownership of their lives . . . But the 'co' of coproduction is now a fallacy; the state won't keep their share of the bargain, most forms of funding and support have gone.

(social worker, 2015)

These extracts suggest something of the pain associated with austerity governance – not only personal pain in seeing services participants had built up disappear, but also the pain of witnessing the increased poverty and social divisions that they saw opening up. Some spoke about the pain of reversals when services they had fought to establish were cut or projects of participation and inclusion became marginalised. Others agonised over the dilemmas they faced as activists struggling to find a voice, and a space for action, in a political climate in which the legitimacy of the state itself was being challenged and feminism, antiracism and other movement struggles demonised:

[As a black woman] it's always been a struggle. In the 90s it was possible to get funding and I worked on community projects for a while. Now there's no money anyway unless it's a crisis. And anyway I'm not sure what I have to offer any more.

(independent trainer, ex-local authority equality worker, 2013)

But some of the most troubling responses reflected a process of revaluing the achievements of the past or agonising about where to place current energy:

Looking back there was a sense of hope – we almost believed that governments could act in our interests. And I wonder if I was right to encourage that. Seeing hope turn to bitterness was really painful.

(local authority worker responsible for community engagement, now redundant, speaking in 2014)

It's all gone – all the things we built up over the years have gone. I'm now part of a group campaigning against the cuts, trying to defend services. I sometimes wonder whether all those critiques we did from the inside [of government bureaucracy] were a good idea – they sort of prepared the way for what's happening now.

(past chief executive, government agency, speaking in 2013)

Earlier I spoke about the ways in which actors with backgrounds in feminist and other social movement struggles had helped generate new governmentalities. These last two extracts hint of the personal agonies at stake as those interviewed reflected back on how far their working lives had helped pave the way for political disaffection and state retrenchment.

This political and ideological climate of austerity governance not only served to legitimise cuts and state retrenchment; it also enabled the surfacing of political rage against the movements that had supported the transformation of state welfare. Feminism, in particular, became the target of dismissal and abuse (see Fraser, 2009; McRobbie, 2009, on the anti-feminist backlash in the US and UK). This is changing the 'affective atmospheres' in which negotiations about governing take place. It also feeds into an anti-statist rhetoric that serves to discredit the institutions that enshrined the equality politics of the twentieth century.

The state, of course, has always been regarded with ambivalence by those struggling for social and political change. Nominally the formal guardians of equality, government bodies have frequently been revealed as complicit in forms of sexual and racial abuse, seeking to protect perpetrators from public blame and shame and to shore up the legitimacy of flawed institutions. This came to attention in the UK in 2014/15 with a series of sexual abuse scandals in which public bodies were shown to have failed to protect vulnerable people in their care. The ambivalent sexual and racial politics of the state was also highlighted in the 2014 Rochdale scandal surrounding the rape and exploitation of young women by mostly 'Asian' taxi drivers,[2] and the 'Trojan Horse' scandal in which members of Muslim communities were charged with attempting to 'take over' the governance of local schools in Birmingham.[3] In each case the institutions of governing (police, local authorities, social services and others) were presented as feeble, incompetent and even cowardly in the mainstream media.

The prevalence of avoidance, repression, blame and shame signals a malign, rather than beneficial, emotional dynamic of governing. This dynamic feeds into the erosion of trust and popular detachment from Westminster politicians and institutions of government, from the EU and from traditional political parties. As well as detachment, some commentators argue that citizen responses to the polity are marked by 'disaffected consent' (Gilbert, 2013: 18), while others suggest there is evidence of increasing popular anger and rage. Jonathan Freedland, commenting on the UK government's proposals to scrap the human rights act in 2014, argued that:

> Anger about excessive powers supposedly wielded by Strasbourg judges, Scottish MPs or the European Union is not really about institutional arrangements. It is instead an outlet for a much more visceral rage, the furious sense that the world is not as it should be – and that someone far away must be to blame. This is the pool of fury that UKIP drinks from, and which the Tories want to channel their way.
>
> (Freedland, 2014: 39)

This 'visceral rage' came to a head in the UK referendum about EU membership in 2016. Rational arguments about the economic benefits of membership of the EU were of little import in the gathering momentum of the 'leave' campaign – a campaign based on summoning up hatred and rage against migrants, against a discredited political elite and against those promoting inclusive, liberal and tolerant values. The referendum process itself was often venomous, and widely acknowledged as promoting a culture of blame, bitterness and division.

Questions of rage, passion, fury may appear to be 'outside' governance. They link the personal to the political, bypassing a public world of formal institutions, laws and policymaking. However, I want to argue that citizen responses to the polity, and governmental attempts to shape citizen feelings, are fundamental to an understanding of a politics of affect. Contemporary politics is characterised by populist right-wing appeals that vilify the social movements associated with an expansive concept of state welfare. A politics of rage is integral to governmental attempts to summon 'the people' as a unified entity attached to a particular representation of the nation. It is also fundamental to understanding the abuse of welfare recipients and others who refuse to comply with notions of responsible citizenship. Such stigmatising practices are intensifying as austerity governance has deepened. Governments, together with a complicit and willing media, have sought to summon up feelings of disgust, rage and indeed hatred for marginal others and to bind 'ordinary', hard working and responsible citizens into a hegemonic 'we' – whose feelings of attachment are strengthened by fear of the other, fear of slipping across the divide that separates.

In such a climate the gap between the dispassionate registers of technocratic governance and the affective registers of political ideology is deeply troubling. It creates the space for the rise of political parties of the right, and for an intensification of populist styles of political rhetoric and action. Austerity governance reaches deeply into the emotional lives of citizens, appropriating and reworking sentiments of community, morality and responsibility – and the movements that generated and sustained them.

Conclusion

This chapter has challenged the idea that emotion governance represents a new, all encompassing governmentality. It has shown how the current policy emphasis on relationships, responsibility and the politics of personal lives derives, in part, from new governmental projects, but also from feminist theory and practice. As such, emotion governance represents the site of active appropriations and negotiations with dominant logics of rule. Rather than assuming a wholesale shift from one regime of governing to another, then, my interest has been in how different governing regimes may be overlaid and entangled, or perhaps conflict, and in the ways the contradictions are managed. This was illustrated through the research cited earlier, and situated in Wetherell's focus on social practice. But rather than referencing these as 'macro' (theories of neoliberalism, the state, governmentality and so on) and 'micro' (ethnographies of practice) I want to hold them together through what I have described as a politics of 'border work'. Border work, in the sense I have used it in this chapter, involves forms of 'affective performance' that unevenly patch together different identifications and allegiances, public and private selves, and personal and political attachments, with considerable psychic strains and discomforts.

And such strains and discomforts are intensifying in conditions of austerity. The political climate described in the final section is one that generates rather different emotional registers of governing from those associated with an expansion of state concern with happiness and well-being – but is inextricably entangled with them. What emerges is not a single logic of governing but different streams or strands that are loosely articulated with each other. The articulations inherent to a politics of austerity governing include:

– A politics of rationality in which practices promoting enhanced governmental attention to feeling, care and well-being are subordinated to an assertive insistence on the primacy of the economy. Politics is reduced to questions of technical competence in 'balancing the books', 'reducing deficits' and prudent economic management. Here governments strive to summon up a public mood of sacrifice, forbearance, patience and the security of a promised return to 'business as usual'.
– A politics of responsibility, negotiated through an uncomfortable negotiation between activist concerns and projects of state retrenchment. Progressive concepts of coproduction, personalisation and community become appropriated and transformed as they confront the governmental promotion of a new moral economy – an economy in which individuals are encouraged to 'feel' responsibility for the well-being and health of the self, the household, the community and indeed the nation.
– A politics of division and rage that pervades and pollutes the 'affective atmospheres' surrounding and reshaping the conditions of such negotiations. We can see new affective registers actively summoned by politicians: the scapegoating of migrants, the demonisation of welfare recipients, and the intensification of the language of security/insecurity in an effort to shore up social control and legitimate governmental intrusion into personal lives.

These three strands – and others not considered here – exist in difficult and problematic alignments, which take different forms in specific locales (nations, regions, localities). They summon up different emotional repertoires and different imaginaries of belonging, morality and responsibility. The ambiguous relationship between them opens up the possibility of the kinds of border work discussed earlier in this chapter. As I write, many actors are struggling to manage the material and emotional consequences of austerity, while also seeking to stitch together creative solutions and new possibilities (New Economics Foundation, 2014). They are exploiting cracks and fissures in dominant governing rationales, and seeking to 'perform new worlds' within the confines of present constraints

(Newman, 2013b). Such worlds are rooted in an attention to human relationships, connections and emotional lives. But the analysis offered in this chapter confounds any notion that 'emotion governance' offers a new positive regime of relational engagement between state and citizen. Equally it cannot be considered as a cynical replacement for well-funded state services. Rather, it is the site of contradictions, tensions and possibilities that are the focus of different forms of 'border work'.

Notes

1 Sure Start was a UK Treasury-funded programme designed to support the development of young children and their families in areas of poverty and disadvantage. Initiated in 1998 under New Labour, it privileged partnership working and involved high levels of local governance autonomy, though these features were compromised in later iterations of the programme.
2 In 2012 there was extensive media coverage of the conviction of nine male taxi drivers for the rape and abuse of teenage girls, together with conspiracy and trafficking charges, in Rochdale, England. The outraged reporting tended to point to the abuse of white girls by men of Pakistani or Afghan origin, reporting that amplified anti-Muslim sentiments and helped stoke anti-immigration political campaigns. But it also highlighted flaws in the responses of social services and other agencies that were accused of avoiding action for fear of being viewed as racist (see Trilling, 2012).
3 This refers to the alleged plot by Muslim groups to infiltrate the governing bodies of selected schools in Birmingham. Four separate inquiries were launched, including those by Birmingham City and the Department of Education, while Ofsted inspections led to the removal of some head teachers and renewed concerns about school governance. Reports in the local paper, the *Birmingham Mail*, featured claims and counter-claims couched in the language of 'feuds', 'lies', 'extremism', 'plots', 'suspensions', 'confrontations', 'rows' and other emotive terms (*Birmingham Mail*, 2014).

References

Anderson, B. (2009) 'Affective atmospheres', *Emotion, Space and Society*, 2(2): 77–8.
Bevir, M. and Rhodes, R.A.W. (2003) *Interpreting British governance*. Routledge, London.
Birmingham Mail (2014) 'Trojan Horse investigation of Birmingham schools', at: www.birminghammail.co.uk/all-abouttrojanhorse.
Brown, B.J. and Baker, S. (2012) *Responsible citizens: individuals, health and policy under neoliberalism*. Anthem Press, London.
Brown, W. (2005) *Edgework: critical essays in knowledge and politics*. Princeton University Press, Princeton, NJ.
Cahn, E.S. (2000) *No more throw-away people: the coproduction imperative*. Essential Books, Washington, DC.
Clarke, J., Newman, J., Smith, N., Vidler, E. and Westmarland, L. (2007) *Creating citizen-consumers: changing publics and changing public services*. Sage, London.
Cook, G. and Muir, R. (eds) (2012) *The relational state*. IPPR, London.
Cooper, D. (2014) *Everyday utopias: the conceptual life of promising spaces*. Duke University Press, Durham, NC.
Davies, K. (2015) *Social work with troubled families*. Jessica Kingsley, London.
Dolan, P. (2014) *Happiness by design*. Hudson Street Press, London.

Fraser, N. (2009) 'Feminism, capitalism and the cunning of history', *New Left Review*, 56: 97–117.

Freedland, J. (2014) 'Scrapping human rights law is an act of displaced fury', *The Guardian*, 3 October 2014, at: www.theguardian.com/commentisfree/2014/oct/03/scrapping-human-rights-law-european-court-ukip (accessed 23 June 2016).

Froggett, L. (2002) *Love, hate and welfare: psychosocial approaches to policy and practice*. Policy Press, Bristol.

Gerhardt, S., Jowell, T. and Stewart-Brown, S. (2011) 'You don't talk about love in government: a roundtable discussion on Sure Start and the first three years of life', *Soundings*, 48: 145–57.

Gilbert, J. (2013) 'What kind of thing is "Neoliberalism"?', *New Formations*, 80–1: 7–22.

Gill, R. and Pratt, A. (2008) 'In the social factory: immaterial labour, precariousness and cultural work', *Theory, Culture and Society*, 25(7/8): 1–30.

Gregg, M. (2011) *Work's intimacy*. Polity Press, Cambridge.

Hajer, M. and Wagenaar, H. (2003) *Deliberative policy analysis: understanding governance in the network society*. Cambridge University Press, Cambridge.

Hochschild, A.R. (1983, 2nd edn 2003) *The managed heart: commercialization of human feeling*. University of California Press, Berkeley.

Hoggett, P. (2000) *Emotional life and the politics of welfare*. Macmillan, Basingstoke.

Holloway, W. (2006) *The capacity to care: gender and ethical subjectivity*. Routledge, London.

Hunter, S. (ed.) (2007) *Coproduction and personalisation in social care*. Jessica Kingsley, London.

Hunter, S. (2015) *Power, politics, emotions: impossible governance?* Routledge, London.

John, P., Cotterill, S., Richardson, L., Moseley, A., Smith, G., Stoker, G. and Wales, C. (2013) *Nudge, nudge, think, think: experimenting with ways to change citizen behaviour*. Bloomsbury, London.

Johns, H. and Ormerod, P. (2007) *Happiness, economics and public policy*. Institute of Economic Affairs, London.

Jones, H. (2015) *Negotiating cohesion, inequality and change: uncomfortable positions in local government*. Policy Press, Bristol.

Jones, R., Pykett, J. and Whitehead, M. (2013) *Changing behaviours: on the rise of the psychological state*. Edward Elgar, Cheltenham.

Kingfisher, C. (2013) *A policy travelogue: tracing welfare reform in Aotearoa/New Zealand and Canada*. Berghahn, New York/Oxford.

Larner, W. (2000) 'Neoliberalism: policy, ideology, governmentality', *Studies in Political Economy*, 63: 5–25.

Layard, R. (2005) *Happiness: lessons from a new science*. Penguin, London.

Le Grand, J. (2006) *Motivation, agency and public policy: of knights, knaves, pawns and queens*. Oxford University Press, Oxford.

Lewis, G. (2000) *'Race', gender, social welfare*. Polity Press, Cambridge.

McRobbbie, A. (2009) *The aftermath of feminism: gender, culture and social change*. Sage, London.

Muehlebach, A. (2012) *The moral neoliberal: welfare and citizenship in Italy*. University of Chicago Press, Chicago, IL.

Needham, C. (2011) *Public services: the personalisation narrative*. Policy Press, Bristol.

New Economics Foundation (2014) *Responses to austerity*. NEF, London.

Newman, J. (2012a) *Working the spaces of power: activism, neoliberalism and gendered labour*. Bloomsbury Academic, London.

Newman, J. (2012b) 'Beyond the deliberative subject? Problems of theory, method and critique in the turn to emotion and affect', *Critical Policy Studies*, 6(4): 465–79.

Newman, J. (2013a) 'Spaces of power: feminism, neoliberalism and gendered labour', *Social Politics*, 20(2): 200–21.

Newman, J. (2013b) 'Performing new worlds: policy, politics and creative labour in hard times', *Policy and Politics*, 41(4): 515–32.

Newman, J. (2014) 'Governing the present: activism, neoliberalism and the problem of power and consent', *Critical Policy Studies*, 8(2): 133–47.

Newman, J. and Clarke, J. (2009) *Publics, politics and power: remaking the public in public services*. Sage, London.

Newman, J. and Tonkens, E. (eds) (2011) *Participation, responsibility and choice: summoning the active citizen in Western Europe*. University of Amsterdam Press, Amsterdam.

Rose, N. (1999) *Powers of freedom: reframing political thought*. Cambridge University Press, Cambridge.

Thaler, R. and Sunstein, C. (2008) *Nudge: improving decisions about health, wealth and happiness*. Yale University Press, London.

Trilling, D. (2012) 'How the Rochdale grooming case exposed British prejudice', *New Statesman*, 15 August 2012.

Twigg, J. (2002) 'The body in social policy', *Journal of Social Policy*, 31(3): 421–39.

Twigg, J., Wolkowitz, C., Cohen, R. and Netteton, S. (2011) *Body work in health and social care: critical themes, new agendas*. Wiley-Blackwell, Oxford.

Wetherell, M. (2012) *Affect and emotion: a new social science understanding*. Sage, London.

Wetherell, M. (2013) 'Feeling rules, atmospheres and affective practice', in Maxwell, C. and Aggleton, P. (eds) *Privilege, agency and affects: understanding the production and effects of action*. Macmillan, Basingstoke, pp. 221–39.

Williams, F. (1999) 'Good enough principles for welfare', *Journal of Social Policy*, 28(4): 667–87.

3 Reframing co-production

Gender, relational academic labour and the university

Bryony Enright, Keri Facer and Wendy Larner

Introduction

In recent years debates about the wider purpose of the university have been dominated by accounts of 'academic capitalism' (Slaughter and Rhoades, 2004). These debates focus attention on the marketisation of the university, new partnerships with industry and government, the growth of entrepreneurial practices, and the increasing quantification of academic outputs. Correspondingly the sociology of education literature is often deeply pessimistic (Deem et al., 2007; Docherty, 2011; Martin, 2011; McGettigan, 2013). In accounts of the so-called 'neoliberal university' attention is consistently drawn to processes of global competitiveness, the ever-increasing encroachment of 'audit culture', and the rise of individualistic academic subjectivities. There is also concern about private sector actors entering the space traditionally occupied by universities and what these changes might mean, particularly for the social sciences, arts and humanities (Shore and McLauchlan, 2012).

In this context, the increasing expectation of UK research councils that research projects should be 'co-produced' with partners beyond academia is understandably treated with some suspicion. Co-production is now a requirement of research funding across the academic spectrum, with both early stage collaboration and ongoing engagement during the research process being privileged in funding proposals. Often this demand for researchers to engage beyond the university is understood as a profoundly instrumental agenda, driven by the challenge of reduced public funding for academic research, and in which academic and public interactions are understood merely as functions of each other. Co-production is read as either the harbinger of the decline of academic freedom and blue skies research or the rise of hyper-exploitative relations in which academics reframe their research to meet the demands of external interests, and 'partners' service the academic need to demonstrate impact for the research.

We argue such accounts are constrained by assumptions about the terrain on which we are working and the nature of the phenomenon being analysed. Our aim is to push accounts of emotional states of governance beyond the realms of marketisation, competition and individualisation that feature in the literature on academic capitalism and neoliberal universities, and beyond the dominant emotional response of resignation, doom or anger (Amin and Thrift, 2013). While we recognise the changing institutional context and funding imperatives, competing aspirations and agendas are at play in the shift towards co-produced research. In addition to the trends noted above, women's studies, indigenous studies, development studies and community development have long been sites from which calls for new relationships between academics and partners have emerged and been enacted (Larner, 2012; Facer, 2014). Insufficient attention has been paid to the potential for co-produced research to facilitate interactions characterised by personal relations of solidarity and reciprocity between activist scholars, and scholar activists, rather than the individualised and instrumental relations of neoliberalism.

Our particular focus is on the gendered dimensions of these interactions. The call for academics to build relations of solidarity with those historically excluded from mainstream research has often been led and/or fundamentally shaped by feminist battles for justice. These battles have been played out in diverse arenas, including healthcare, sociology, prisons, neighbourhood renewal and social services provision, and by major disciplinary figures, including, amongst others, social psychology (Michelle Fine), geography (J.-K. Gibson-Graham), sociology (Raewyn Connell), education (Pat Thomson, Linda Tuhiwai Smith) and health and child services (Angie Hart). In these research traditions, the demand to build external relationships operates in the name of various publics (Martin, 2011). It is not premised on instrumental relations between academics and external 'stakeholders', but on deep interpersonal relations which aim to produce socially significant knowledge based upon a reciprocal understanding of the expertise and insights of both partners in the research process (Torres and Reyes, 2011; Eikelund, 2012).

Seen from this perspective, the current injunction to 'co-produce' research might be usefully seen as an opportunity for feminist academics and/or those influenced by wider debates about inclusion to leverage their long-standing political commitments and experiences in diverse organisational settings. In particular, it may be an opportunity to further alter the emotional terrain of research which has been historically shaped by patriarchal and positivist approaches that privilege dispassionate and detached methods (Anderson and Smith, 2001; Greco and Stenner, 2008). Whereas traditional research practice encourages clearly defined academic identities and professional relationships, co-produced research requires reflexive engagements and emotional labour. The focus of research management shifts from defining functional relationships to managing interpersonal dynamics. It also marks a move away from the presumed masculinity of traditional leadership styles valued in the academy (the lone scholar, the globally mobile intelligentsia). In short, the rise of research co-production is not reducible to the neoliberalisation of the university – although this remains an important frame – but is also a chapter in the history of women, feminism and gendered labour in the academy.

We are not the first to identify the 'anatomy of intimacy' that describes the demands that saturate the time and labour of scholars committed to counterhegemonic teaching and research (Berlant, 1997). Feminist scholars and students have long aspired to greater collaboration and cohabitation across different identities, knowledges, struggles and ways of knowing and being inside and outside of the university. In this chapter, we identify the different emotional registers emerging from the struggles between institutional demands of the academy and aspirations for collaborative research. We address the entanglements of this dynamic – including feelings of excitement, passion, anxiety, trust and responsibility – showing that they cannot be separated from the relationships that underpin them.

Multiple dynamics

Others have drawn attention to new ways of working that arise from the conjuncture between change in women's working lives and political commitments (Newman, 2012). There are multiple dynamics at play as more women move into leadership roles and juggle professional, political and ethical commitments. For example, the long-standing literature on so-called 'femocrats' (Yeatman, 1990) examines the experiences of feminist women who find themselves in leadership positions in national and international institutions, including universities. A cognate literature focuses on 'state feminism', the widespread emergence of bureaucratic structures that deal with women's policy issues or gender equality (Haussman and Sauer, 2007; Mazur and McBride, 2007). These accounts make it clear that both women and feminism are well and truly inside public institutions and mainstream organisations, even if the under-representation of women remains and this shift is not always explicitly named as feminism (Walby, 2011).

As collaborative governance, partnership working and multi-stakeholder arrangements have proliferated, a cognate literature has examined how feminism has shaped forms of leadership, working lives and political practice (Alvarez, 1999; Bondi and Laurie, 2005). The increasing importance of 'relational expertise' (Edwards, 2013) and the distinctive emotional and cognitive capabilities required to facilitate inter-professional and inter-agency relationships have been highlighted (Larner and Craig, 2005; Freeman, 2009). Analysts have begun to examine the experiences of women and other traditionally excluded groups who are re-shaping policies and practices in a wide range of settings and inventing new organisational forms (Kantola and Squires, 2012; McRobbie, 2011). Indeed it could be argued that the increasing visibility of interstitial women 'working the spaces of power' (Newman, 2012) is one of the most striking consequences of the conjuncture between neoliberalisation, new social movements and changing patterns of employment and labour.

To date, little is known about how these wider trends are reshaping the university and contemporary knowledge forms. As the academy was democratised and massified in the 1980s, more women were able to take up opportunities for tertiary education and then move into academic careers. There is evidence from earlier studies that feminist academics have acted as change agents in universities (Deem and Ozga, 2000; Morley, 1999). In turn, these shifts are explicitly linked to new managerial approaches, including facilitative leadership styles, soft skills, networking, the presentation of self, the mainstreaming of equality and diversity strategies and so on. While we recognise the intractability of gendered inequalities within and across the academy, we are interested in the extent to which research co-production might be seen as part of an ongoing process emerging from the greater visibility of academic women, their political commitments and organisational reconfigurations.

From this perspective, we ask whether the 'co-production imperative' is providing an opportunity for a new form of relational academic labour that exploits the democratising and pluralising potential of collaborative research, while still continuing to navigate the path dependencies of gendered academic institutions, including ideas of what counts as 'good' research and a 'successful' academic career (Collini, 2011; Calhoun, 2011). There is a great deal at stake here given the appeal of co-produced research for new generations of academic women, including those from traditionally marginalised groups. In this context we wish to understand how senior academic women have responded to the imperative for co-production, what sort of labour is involved in this work, and how they negotiate the competing imperatives of community engagement and established institutional requirements. Finally, and in relation to the specific theme of this book, we want to know whether the emotional landscape of the academy is being reconfigured through the changing infrastructures and imperatives of co-produced research.

Women, leadership and Connected Communities

Our case is the UK Arts and Humanities Research Council (AHRC) Connected Communities (CC) research programme, and the senior women who are using this programme to advance their ambitions for a more inclusive, engaged university. We focus on this programme for two reasons. First, the CC programme is explicitly designed to promote collaborative research into 'community' and is a site in which the lived experience of co-production can be examined in multiple disciplinary and institutional settings. Second, we are closely connected with this programme, as Post-Doc researcher (Enright), as Leadership Fellow (Facer) and as Co-investigator on a large grant (Larner). Consequently we have much at stake in examining whether this programme has the potential to operate as a democratising moment.

Connected Communities is one of a number of large research council programmes which have, over the last decade, explicitly set out to promote strong links between academics and external partners. The programme itself is a site through which competing interpretations of the co-production imperative have been negotiated. Its initial launch was greeted with scepticism by many in the academic community, coming as it did alongside the newly elected UK government's announcement of the 'Big Society' programme (read by many as an opportunity to withdraw funding from the already beleaguered community and voluntary sector). At the same time, those with backgrounds in community and engaged research were quick to identify the CC programme as an opportunity to promote and fund more deeply participatory research methods. Participants in early workshops, for example, successfully made the case for stronger community representation in commissioning and delivery of research, for a more complex and nuanced view of communities, and for greater awareness of the potential benefits of collaborative research for both academic and public participants. Co-production has emerged as the concept that captures these ambitions, and is now becoming increasingly prevalent across the research terrain.

Since its establishment the CC programme has spearheaded a number of innovations in research arrangements. These include: enabling community partners to be named as co-investigators in the research, engaging community representatives in Research Development Workshops, supporting new two-stage funding models, building strategic partnerships that allow community organisations to have control over their own proportion of the funding, recognising the plurality of research outputs that emerge from collaborative research, and funding community partners to build research relationships and co-design projects. Seen collectively, these new ambitions and expectations have transformed the wider research landscape and are giving rise to what we will call herein 'relational academic labour'.

By January 2015 the CC programme had funded 314 projects ranging from small-scale scoping studies to multi-million pound five-year programmes. Within the programme women academics have been relatively successful, and there are 151 female Principal Investigators (PIs) and 163 male PIs at the time of writing (May 2015). Because of the programme's aim to build a cohort of researchers, many of these have been supported to lead on more than one project, with 34 men and 32 women serving as PI more than once. However, despite this seeming gender balance, there are also familiar patterns that emerge on closer investigation. Across the ten high-status large grants on the programme – these projects range from £1.1 to £1.9 million – there are seven male PIs and only three female PIs. There are also some concerns about the ability of arts-led programmes to address structural inequalities in marginalised communities, and the potentially exclusionary nature of the funding processes discussed in more detail below. However, for all participants the CC programme has become a key site for co-produced research practice, and made visible the opportunities and tensions that materialise in collaborative research settings.

The discussion that follows is based on 12 in-depth interviews conducted in 2014 with women who have played leadership roles in the programme: as Principal Investigators of their own projects, as influential participants in workshops and events, or as Co-investigators leading a significant programme of work on large grants. They are all employed as Senior Lecturers or above in their own institutions, a factor that is important in explaining how they navigate the complexities of leading co-produced research in universities that are still geared up for a different research model. They also represent different disciplinary traditions and academic trajectories.[1]

We are particularly interested in identifying the new emotional registers that emerge in the co-production of research. To do so we focus on three key moments of co-production practice. These moments are powered by and themselves engender new practices of relational academic labour, namely the intertwined emotional and cognitive labour required to conduct research that is embedded in an ethic of reciprocity and shared purpose:

- first, becoming a leader of co-produced research;
- second, inventing new research temporalities;
- third, doing good research and being a good researcher.

Leading co-produced research

The 12 women who are leaders in the CC programme identify themselves variously as ethnographers, philosophers, medical researchers, designers, historians, lawyers and geographers. More importantly for our analysis, the vast majority have biographies that offer them multifaceted identities allied to external political and social commitments. One PI brings a history of working in adult literacy, another a history of community development and school leadership; both of them entered postgraduate study as mature students and have developed academic careers that built upon and benefited from that community-based experience. Others have more traditional academic trajectories but are shaped by political orientations that have led them to build relationships between academic practice and traditionally excluded groups, either through advocacy research or through policy-engaged evaluation and research. Two of our interviewees, both historians, were participating in collaborative research for the first time, although both have a track record of civil society engagement in their personal lives. The CC programme allowed them to bring their traditional archive-based research into dialogue with personal political and community-engaged identities.

These women exemplify the forms of feminist-inflected activism that Newman (2012) discusses, not necessarily explicitly identifying as either 'feminist' or 'activist' but embodying, living and performing a range of alternative commitments and practices through their work. The CC programme was an opportunity for them to be recognised and resourced to do research that had, historically, been conducted at the margins of UK universities. For one academic, the opportunity was greeted with sheer delight:

> This is the thing that I have been waiting for all my life . . . it was like a fantastic moment because I just thought 'Wow'. And it was everything that I wanted to do in my career, because it was community plus arts and humanities – which I'd struggled with in other sort of bits of my job.
>
> (Christine, 2014, Professor)

All the women interviewed were explicit about the ways in which their prior experiences and relationships, and the political and community engagements that were integral to these, had prepared them for the research practice that the CC programme subsequently validated, often for the first time.

These background experiences manifested themselves in an enormous amount of optimism, energy and excitement about the CC programme. The imperative to co-produce research was experienced not as a neoliberal incursion on academic autonomy, but as an opportunity to conduct socially engaged, politically significant, academic research they had previously struggled to conduct in other ways.

Indeed, one interviewee observed that research council funding potentially liberated them from other powerful vested interests:

> We'd done a lot of work with policy makers that always sought to question what they were doing and ask them to reflect on the extent to which their policies and actions impacted on different communities but didn't directly engage with those communities themselves. And that was partly because the funding opportunities weren't there ... So when Connected Communities came along it was very interesting for those reasons.
>
> (Brenda, 2014, Reader)

Critically, however, the co-production moment was less a novel incursion into a new research practice than an opportunity to enhance already existing academic, civic and personal commitments. The emotional labour of working with collaborating organisations was, in the main, already in train and the relationships in place.

What was new was the relational academic labour involved in the complex process of gaining funding, in particular building the coalitions and networks of collaborators needed to successfully navigate research council and peer review processes. Co-production has emerged alongside a set of wider innovations in which interdisciplinarity, alliance building and the rapid exchange of ideas are foregrounded, along with associated technologies such as sandpits, academic 'speed dating' and novel forms of facilitation (Barry and Born, 2013). The CC programme exemplifies this shift, with a significant element of funding awarded to those who have first participated either in 'Summits' or 'Research Development Workshops'. These invitation-only events are characterised by a set of ideational techniques where 'themes' and 'topics' for research, which fit under the predefined subject matter of the funding call, are identified by participants. Consensus techniques such as Open Space approaches, which rely on self-organisation, free movement of individuals between different groups and ideas, and a pattern of flow between short plenary-style presentations and group work, are used to encourage groups to form around emerging ideas. Research Development Workshops tend to conclude with opportunities to 'pitch' for proposals, with some projects given permission to continue and others shut down. The subsequent research design involves project leaders building teams, (re)negotiating participation, budget and resource allocations, and shepherding proposals through complex review processes, sometimes lasting over two years from the first workshop.

This bid development process involves significant amounts of emotional and cognitive labour. Participants have to negotiate assumed expectations of 'research excellence' that drive the funding process, as well as the complex terrain of interpersonal relations and emotional commitments that arise in the workshop. Many interviewees describe the workshops as highly combative spaces, in which 'generating collaboration' is a veneer covering a highly competitive and individualised process. Success in such settings requires the ability to engage with and identify points of common interest and complementary expertise, as well as act in a deeply self-interested manner. Arguably, the workshops risk encouraging a highly instrumental view of fellow collaborators; community partners report being sought out as 'bid candy', while some academics reported anxieties about the behaviours that it produces in them:

> I came away thinking 'three days, no sleep, totally mental – I don't know that that brought out the best in me'. Cos you know it encourages a competitive like 'Who's doing that over there?' you know. And afterwards I thought 'Ooh I'm not sure I'm very proud of myself'. But having said that, I came out of it as part of a £1.5 million project.
>
> (Brenda, 2014, Reader)

One interviewee conjectured that her lack of success in the large grant process was partly related to the difficulties of negotiating the competing emotional and intellectual demands of this bidding process. Having built personal relations with colleagues in the workshop setting, which included a degree of commitment to inclusion and openness to all participants' ideas, she was not prepared to 'streamline' the group to achieve the familiar coherent and focused proposal needed for success in the peer review process. Importantly, however, she also noted that this moment of 'failure' had been one of the most intellectually rewarding experiences of her career as it had laid the foundation for longer-term collaborations. Another interviewee observed that the process of 'culling' team members with whom relationships had been formed in the workshop was particularly difficult: 'We made a few decisions about bringing extra people in. But then we had to make a decision about letting one go as well, which was horrible' (Gemma, 2014, Senior Research Fellow).

More recent participants report handling these competing imperatives in different ways, by making explicit the 'rules of the game', and engaging all colla-borators in evaluating their contributions against external criteria. In these accounts the participants recognise the explicitly instrumental nature of the competitive bidding process, without reducing interpersonal relations to such instrumentality. The extent to which this tactic offers an opportunity for truly collaborative working, a temporary expediency that enables exploitation of a funding opportunity, or a simple co-option of the messy and multiple agendas of community collaborators to the disciplinary practices of the academy, remains up for debate. More generally, however, what such tactics make manifest is the extent to which personal and ethical relationships are now an explicit part of this research terrain, and the increasing codification of relational academic labour as an example of governing emotional states in a highly charged context.

Success in this field also engenders new hierarchies and power relations, which in turn demands further relational academic labour. Indeed success is itself often a new experience for scholars committed to participatory research, many of whom are used to occupying marginal disciplinary positions. As the PI of one large grant explains, 'I had never received research council funding for participatory research before' (Stephanie, 2014, Professor). Consequently a core element of success is negotiating what it means to be identified with the university as a dominant and powerful social institution.

> I think it's sometimes about being honest about what it is we're doing and what we're not doing and the role of the University, but without undermining the fact that we're really privileged. It's really difficult the privilege and the inequalities . . . you know we are fundamentally fully time paid to do this, and ultimately we will go . . . we'll try not to go away at the end of the day, but we probably will. So there's something about honesty and expectations. There's something about validation and visibility.
>
> (Brenda, 2014, Reader)

Much of the emotional labour lies in recognising and, over time, attempting to resist what Berlant (1997) calls the 'fantasy of rescue and identity' of the politically engaged academic. By this she refers to the intimacy of learning and pedagogy between academics and their students, and the increasing responsibility on teachers to support multiple aspects of their students' development, both academic and personal. In CC we see these senior female academics attempt to produce a more measured and informed understanding of the material and intellectual conditions that underpin and constrain co-produced research, while at the same time confronting the intimacy of this work and the limits of what differently positioned individuals can do for one another.

Inventing new research temporalities

Co-produced research requires reconfiguring the temporality of conventional academic subjectivities. By this we mean that co-produced research projects are understood, first, as part of a longer set of pre-existing and ongoing relationships; second, as requiring time and commitment that extend beyond the limits of formally paid academic labour; and, finally, as a shared endeavour, a process of working towards shared goals that go beyond the time-frames of the individual project.

The length of relationships, time spent with people and commitment to ongoing endeavours, to seeing things through, came out very strongly in the interviews. Long-term relationships are a necessary part of proving credibility, building trust and demonstrating a commitment to work with communities. The visible valuing of these long-term relationships comes with its own implications for emotional labour, responsibility and care. These relationships can be a source of strength, allowing for tensions and disagreements to be confronted and dealt with without disrupting projects: 'When you're working with people you have already worked with, you know what to do when you hit hiccups' (Rita, 2014, Senior Lecturer). They also protect against the idea of an exploitative and extractive research project: 'If you're going to do anything . . . you don't do it as part of a quick in, quick out project. You set up long-term relationships and you're in it for the long haul' (Maureen, 2014, Professor). Finally, and not surprisingly, relational academic labour can become personal friendship:

I don't think there is a way to separate that [professional and personal relationships]. I think because of the rhythms and the temporalities of community organisations I think, and in order to maintain and manage relationships, you have to be prepared to have it be part of your life . . . there's no boundary.

(Rita, 2014, Senior Lecturer)

Because of this wider commitment, large amounts of unpaid labour become key to doing co-produced research effectively. 'This is a whole life thing, and you can't just clock off at 6pm' (Rita, 2014, Senior Lecturer). Of particular note are the ways in which the community/evening work that is necessary to building the relationships brings incursions into family commitments or, if family is prioritised, then guilt can ensue: 'I mean you know I still have a teaching load, still have my you know REF stuff to do. So I just couldn't spend every evening, and I've got young children, I couldn't spend every evening going to these community meetings' (Sally, 2014, Professor). This commitment to the wider relationship also means juggling multiple temporalities:

The time I have written into the projects and the time that I spend on the projects has no relation to each other. But I've just accepted that you know, I mean that's true to academia. So you know you . . . you make your own bed in a way, I think. I think you can't be too bean counting about these things, if you're going to do it you have to do it. And you have to protect yourself a little bit so you don't go mad, but other than that you can't pretend that it has a real relationship [to the time written in the bid] because it doesn't.

(Sally, 2014, Professor)

Finally, the long-term, continuing relationships can be difficult to maintain under the institutional pressure of the university; in some cases they required academics to work for free and maintain relationships outside their work until they were able to find new projects and grants which could bring these communities back into a formal (funded) research partnership.

Developing this kind of research relationship changes the nature of research practice. The relationships are not something these women can or want to switch off from, in terms of either the emotional and social ties to the people or reproducing boundaries between their research and personal life. Consequently they push against conventional understandings of academic labour:

I would probably be looking for longer term kind of thinking, to try and get past the project kind of mentality. I don't know how that would . . . I can see there's lots of dangers around that, but I think I might be looking for more evidence that this is not, kind of, just an opportunist way to get a bit of money, but there are people that are committed to the long term effort.

(Maureen, 2014, Professor)

This emphasis on relationships contrasts with institutionally entrenched, masculinist, academic subjectivities that emphasise individual ambitions and economic status. The focus on collaboration, friendship and reciprocal trust means the research is motivated by the need to achieve progress for multiple actors within and beyond the university, rather than the aspirations of an individual academic. Co-produced research allows academics to do research as part of a 'shared path', rather than a personal career trajectory. It creates duties of care, manifest in ongoing attempts to make the research relationship more reciprocal and to recognise the multiple benefits of the projects. This duty of care can also take practical forms, for example leveraging the university as a resource and creating connections between partners and other university facilities and networks. None of this fits easily into the compartmentalised frameworks and individualised processes of the contemporary university.

These orientations to research – personal, committed, engaged, reciprocal, engendering duties of care, temporally flexible – are recognisable as part of the 'intimacy expectation' (Berlant, 1997) of feminist academic labour. Such orientations also bring risks of stress, anxiety and, at times, fractured relationships when they confront conventional conceptions of academic labour. They have the potential to open up the forms of flexible, always-on labour which, rather than being a powerful countervailing force in the neoliberal university, risk the intensification of self-exploitation upon which the new university depends. One of the questions we need to better understand, therefore, is how and when such temporal disruptions become mechanisms for social relationships that can build new power bases to contest human instrumentalisation, and how and when they become part of the intensification and flexibilisation of labour in the academy.

Good research and the good researcher

Co-produced research – as a practice of reciprocity rather than instrumentality – brings with it fundamental questions about what counts as 'good' research and a 'good' researcher. For some of the women interviewed, these projects catapulted them into the limelight as successful and high-profile researchers, having achieved significant funding success and being able to contribute to a new narrative of the university as a socially engaged institution. Such a change in status was experienced as a situation that was to be variously welcomed, questioned and resisted:

> I think that sometimes the University, you know they may be like having a photo of academics working with different sections of the community, but actually what really matters is the bottom line and the fact that it's got a research income behind . . . that's why I get a bit cynical because I think it's sometimes a bit superficial maybe, because they like the picture.
>
> (Sally, 2014, Professor)

The 'good' researcher in this institution is clearly one who produces not only research income but also great photo opportunities. However, in co-produced research the 'good' researcher also has an allegiance to her collaborating partners, something that brings them into tension with institutional requirements at times, as the same researcher observes:

> [The university press office] rang me up about it and they said they weren't very happy with it because they thought that they really wanted some different coloured faces there . . . I said well actually if you'd come to my [other] event we have a big ethnic mix . . . But I just thought oh it's just ticking a box . . . it's a bit like politicians setting up photo shoots.
>
> (Sally, 2014, Professor)

The need to perform to institutional expectations is joined by disciplinary measures of value, where the spectre of 'REF' and 'REFable' outputs looms large. This challenge is familiar from previous analyses of the implications of engaged and collaborative research in an academy that operates using text-based journal papers as a measure of quality (Leathwood and Read, 2013). Meeting the increased expectations for research excellence becomes increasingly challenging in a context where there are counter-claims for jargon-free, accessible publications that speak to a wider audience, and sometimes even calls for collaborative writing as well as co-produced research design.

More interesting, perhaps, is the way in which these academics were themselves negotiating and regulating discussions of what counted as research itself, let alone 'good' research. In the context of 'austerity politics', many community organisations are seeking new sources of funding for ongoing activities, and there is often slippage between their wider ambitions and the goals of the research project. Moreover, external measures of quality that are often the subject of critical concern are, in fact, deeply internalised in the identities, assumptions and practices of the academics themselves:

> I actually made a slide just today for the [community collaborator] researchers, just to try and like separate 'this is what research is, and this is what you guys have probably been doing so far' – it's not business as usual. We're not looking at your existing services, we're doing something different.
>
> (Gemma, 2014, Senior Research Fellow)

Here the significance of the longer-term temporality of these partnerships becomes visible, as the ongoing negotiation of what counts as 'good research' and what is meaningful to collaborating partners is explored and developed over a series of projects. One of the most successful academics in the programme described the tentative, lengthy, provisional and negotiated process by which these public and academic relationships and conversations are formed into tractable research ideas that may satisfy the competing demands of the discipline and a wider social purpose:

> I remember sitting in the community centre with [a youth worker] and the young people and him coming up with ideas . . . It really was from that

bottom up perspective. And for that I felt confident enough to PI that just because the project involved two [institutions], which I had links with, the youth service which I had links with, a community organiser person . . . who I had links with, an artist . . . who I was working with on the other projects, the [academics] – and . . . we brought in a poet. So I sort of knew . . . I felt I could choreograph everybody.

(Christine, 2014, Professor)

Being a 'good researcher' with this dual perspective on the academy and the partnerships with collaborators has costs. Institutional allegiances encourage ongoing bid writing, partners who have built deep and trusting relations can see meaningful opportunities to collaborate, and this can push in the direction of continuing what one academic calls the 'treadmill' of project success:

The heart-breaking thing is, it's quite sad this, we then came up with an amazing project, she [one of the community partners] said, 'Okay why have you said you won't do it, because I want to do something with you . . . ?' I said, 'No no no no, I've got to stop, I've got to stop applying for money.' But I think I am right not to apply for money, but I think it's that thing by which if somebody trusts you like with an experience of racism, which actually, as a kind of white middle-class person, and she's from the British Asian community – that's quite special . . . that's what's so sad about [not being able to make another bid].

(Christine, 2014, Professor)

Here the limits of the reciprocal, friendship-based model of academic–public collaboration are perhaps most visible. As the university itself is unable to hold and sustain these relationships when the energy and intimate labour of the individual academic is depleted, this can affect the relationship and trust built up between academic and community partner. Charisma, friendship, personal commitment are necessarily finite resources embodied in the individual academic. If research co-production is to diversify the knowledge production practices of the university, if the feminist ambitions for reconfiguring the emotional landscape of the academy are to be achieved, then the broader structural conditions that cause collaborations between civil society, community groups and other marginalised institutions to be premised upon the fragile thread of personal commitment will need to be addressed.

Conclusion

Critiques of co-production as the marketisation and instrumentalisation of the academy are clearly insufficient to explain the motives, practices and consequences of the relational academic labour these academics have achieved through the CC programme. However, research relations based on personal reciprocity, charismatic leadership and unbounded commitment are also unsustainable for individual academics and the more long-term project of transforming the emotional and intellectual terrain of the academy. Such narratives risk obscuring the extent to which academics in these positions are *both* marginalised and powerful, *both* subject to institutional pressures and operating with a degree of intellectual freedom that is unfamiliar beyond the academy.

What is clear, however, is that these academics are aware of and working through these contradictions and tensions through their ongoing research, with some notable successes and (fewer) difficult failures. They are developing a new form of relational academic labour that contests the narrow sensibilities of accepted research practice, and privileges collaboration and co-production. The question that remains is whether such a working through will happen sufficiently quickly to evolve the new institutional forms, accountability metrics and interpersonal relationships to enable the 'co-production moment' to be harnessed successfully as a resource for new forms of public academic labour. One of the interviewees observed:

> I felt this was something like the last push . . . you know if you couldn't do it with this programme, then when would you be able to actually do it, and to what extent does the weight of the institution eventually bear down upon you.
> (Brenda, 2014, Reader)

While we cannot predict the outcome of this process, what we have argued in this account of these 12 academics' practices is that the co-production moment is a further chapter in the history of women, feminism and gendered labour in the academy. Co-production is being appropriated and led by academic women who explicitly understand themselves as generating novel forms of research practice. They are using their established community relationships, their interpersonal skills and significant amounts of emotional labour to negotiate not just with their community partners but also with the rapidly changing expectations of research councils and universities who now see co-production as a means of delivering on wider ambitions for stakeholder engagement. To return to where we started, it is through this relational academic labour that these women are making visible alternative ways of performing the university, and contributing to the wider questioning of who research is for and what the future of the university should be.

Notes

1 We are aware, in focusing on women alone, that we are ignoring the extent to which men's work in the academy may also be changing. We are also aware that the experiences of women in the collaborating community organisations merit greater analysis. These are important considerations and will be the subject of future writing, but for the purposes of this chapter we are explicitly interested in senior academic women's experiences of relational academic labour.

References

Alvarez, S. (1999) 'Advocating feminism: the Latin American NGO "boom"', *International Feminist Journal of Politics*, 1(2): 181–209.

Amin, A. and Thrift, N. (2013) *Acts of the political: new openings for the left.* Duke University Press, Durham, NC.

Anderson, K. and Smith, S. (2001) 'Emotional geographies', *Transactions of the Institute of British Geographers*, 26(1): 7–10.

Barry, A. and Born, G. (eds) (2013) *Interdisciplinarity: reconfigurations of the social and natural sciences.* Routledge, London.

Berlant, L. (1997) 'Feminism and the institutions of intimacy', in Kaplan, A. and Levine, G. (eds) *The politics of research.* Rutgers University Press, New Brunswick, NJ, pp. 119–34.

Bondi, L. and Laurie, N. (eds) (2005) *Working the spaces of neoliberalism.* Blackwell, London.

Calhoun, C. (2011) 'The public mission of the research university', in Rhoten, D. and Calhoun, C. (eds) *Knowledge matters: the public mission of the research university.* Colombia University Press, New York, pp. 1–33.

Collini, S. (2011) *What are universities for?* Penguin, London.

Deem, R. and Ozga, J. (2000) 'Transforming post-secondary education? Femocrats at work in the academy', *Women's Studies International Forum*, 25(2): 153–66.

Deem, R., Hillyard, S. and Reed, M. (2007) *Knowledge, higher education, and the new managerialism: the changing management of UK universities.* Oxford University Press, Oxford.

Docherty, T. (2011) *For the university: democracy and the future of the institution.* A&C Black, London.

Edwards, A. (2013) *Being an expert professional practitioner: the relational turn in expertise.* Springer, London.

Eikelund, O. (2012) 'Action research – applied research, intervention research, collaborative research, practitioner research, or praxis research?', *International Journal of Action Research*, 8(1): 9–44.

Facer, K. (2014) 'Más allá de la desagregación y el elitismo: Un nuevo futuro para la universidad autónoma popular?' [Between disaggregation and elitism: a new future for the popular autonomous university?], in Gewerc, A. (coord.) *Conocimiento tecnología y enseñanza: políticas y prácticas universitarias.* Grao, Barcelona.

Freeman, R. (2009) 'What is translation?', *Evidence & Policy: A Journal of Research, Debate and Practice*, 5(4): 429–47.

Greco, M. and Stenner, P. (eds) (2008) *Emotions: a social science reader.* Routledge, London.

Haussman, M. and Sauer, B. (eds) (2007) *Gendering the state in the age of globalization: women's movements and state feminism in postindustrial democracies.* Rowman and Littlefield, Boulder, CO.

Kantola, K. and Squires, J. (2012) 'From state feminism to market feminism?', *International Political Science Review*, 33(4): 382–400.

Larner, W. (2012) 'Beyond commercialisation', *Social Anthropology/Anthropologie Sociale*, 20(3): 287–89.

Larner, W. and Craig, D. (2005) 'After neoliberalism? Community activism and local partnerships in Aotearoa New Zealand', *Antipode*, 37(3): 402–24.

Leathwood, C. and Read, B. (2013) 'Research policy and academic performativity: compliance, contestation and complicity', *Studies in Higher Education*, 38(8): 1162–74.

McGettigan, A. (2013) *The great university gamble: money, markets and the future of higher education*. Pluto Press, London.

McRobbie, A. (2011) 'Re-thinking creative economy as radical social enterprise', *Variant*, 41, at: www.variant.org.uk/ (accessed 12 December 2015).

Martin, R. (2011) *Under new management: universities, administrative labour and the professional turn*. Temple University Press, Philadelphia, PA.

Mazur, A. and McBride, D. (2007) 'State feminism since the 1980s: from loose notion to operationalised concept', *Politics and Gender*, 3(4): 501–13.

Morley, L. (1999) *Organising feminisms: the micropolitics of the academy*. Macmillan, London.

Newman, J. (2012) *Working the spaces of power: activism, neoliberalism and gendered labour*. Bloomsbury, London.

Shore, C. and McLauchlan, L. (2012) '"Third mission" activities, commercialisation and academic entrepreneurs', *Social Anthropology/Anthropologie Sociale*, 20(3): 267–86.

Slaughter, S. and Rhoades, G. (2004) *Academic capitalism and the new economy: markets, state, and higher education*. Johns Hopkins University Press, Baltimore, MD.

Torres, M. and Reyes, L. (2011) *Research as praxis: democratising education epistemologies*. Peter Lang, New York.

Walby, S. (2011) 'Is the knowledge society gendered?', *Gender, Work and Organisation*, 18(1): 1–29.

Yeatman, A. (1990) *Bureaucrats, technocrats and femocrats*. Allen and Unwin, Sydney.

Part II

Emotions in public policy-making

4 Choice architecture as new governance

The case of the Dutch housing market

Kayleigh van Oorschot, Menno Fenger and Mark van Twist

Introduction

In their attempts to fight complex social problems, policy-makers are vulnerable to what could be called 'the rationality trap': assuming citizens behave rationally whereas in reality they do not – or at least not all the time. In the rational policy model (Bovens et al., 2012), three types of tools can be used to guide citizens' behaviour: (1) judicial tools (commandments and prohibitions); (2) economic tools (financial incentives); (3) communicative tools (persuasion and warning). These tools 'assume' that the subjects of the policy are driven by rational motivations. In other words, classical policy instruments presuppose that people's actions aim towards deliberatively articulated goals that are more or less stable. The emotional state of individuals, however, can be considered as one of the many reasons why people do not act in accordance with the rational model. This chapter explores the role of emotions and non-rational behaviour in a domain that until now has primarily been analysed from a rational, economic perspective, namely the housing market (see for an exception Christie et al., 2008).

The housing market in the Netherlands, just as in many other countries, came almost to a standstill following the financial and economic crisis that started in 2008. In response, and rather innovatively, the Dutch Ministry of Housing asked us to help them rethink the role of 'non-rational' behaviour to reintroduce dynamics into the housing market, thereby acknowledging 'all kinds of emotions' (Christie et al., 2008: 2297) that guide the behaviour of buyers and sellers. This chapter builds upon the insights from this research and critically reflects upon the results.

Christie et al. (2008: 2298–9) show that non-rational – or emotional – behaviour in the housing market is connected to two causes. First, it seems that certain emotions (most notably those of confidence, optimism and greed) drive up the prices paid for properties. Second, the home is considered as an emotional space and an expression of identity, rather than as a 'bunch of stones' (our words). These emotional causes for behaviour require non-rational policy responses. Therefore, a starting point in our advice on the Dutch housing market was found in the work of Thaler and Sunstein (2008) on 'nudging' (see Pykett et al., Chapter 5 in this volume). Through the concept of nudging we explore how 'non-rational' modes of governance may be applied to unfreeze the Dutch housing market, and what some of the challenges and limitations of nudge might be.

We have divided this chapter into four sections. First of all we present the background to nudge and explain what a nudge is, and what types of policy tools different nudges offer. Next we discuss how these tools might be applied to the Dutch housing market. In this section we by no means attempt to cover the entire field of the housing market or to present ready-for-use policy proposals. Rather the goal is to illustrate the difficulties that come with applying nudge, its ambiguity and the deliberations that have to be addressed. The third section offers an overview of the dilemmas that we have encountered when considering how to apply nudges to the housing market. These dilemmas are quite general and show that we are still at the beginning of the discussion on nudges and their application. Finally, in the concluding section we reflect on the possibilities and limitations of governing non-rational and emotional behaviour.

Studying human behaviour: backgrounds

In their 2008 book *Nudge: improving decisions about health, wealth and happiness* Richard Thaler and Cass Sunstein offer an alternative perspective on policy-making based on insights from psychology and the behavioural and social sciences. As others outline, these disciplines demonstrate that most of the time we do not act rationally at all:

> Behavioural sciences have enabled policy-makers to move beyond narrowly conceived understandings of human decision-making that assume considered and rational responses ... to the choices we are confronted with. Those designing policy can now draw on research that reveals the role of emotion, social context, automatic response, mental shortcuts, and intuition within human behaviour.
>
> (Whitehead et al., 2014: 11)

Human behaviour is thus, according to Thaler and Sunstein, influenced by 'non-rational' elements. These non-rational elements were not original ideas of Thaler and Sunstein; rather the latter were influenced by numerous scientific and social scientific approaches. In this section we focus on the origins of research approaches to three of these non-rational elements: (1) automatic response; (2) social context; and (3) emotions, mental shortcuts and intuition.

First, in terms of automatic response, in around 1913[1] behaviourism emerged in reaction to a speculative focus on the mind and consciousness by philosophers and the first psychologists. Adherents proclaimed psychology as an experimental and objective natural science that took observable behaviour as the only valid object of psychological studies. Any reference to non-observable consciousness and mental attitudes would therefore be banned (Jones and Elcock, 2001: 49). Exemplary for this development were the experiments by Ivan Pavlov involving the stimulus–response relation in dogs: he found that dogs that were repeatedly confronted with a stimulus followed by food started producing saliva after several of these sequences merely in response to the stimulus (Hothersall, 1995: 479). Behaviourism focused mainly on stimulus–response. In nudge theory, similarly, many decisions made by people are supposed to be driven not by rational deliberation, but more by an automatic response.

Second, 'social context' originated in social psychology, which studies the nature and causes of individual behaviour in social situations. The roots of this discipline go back to 1898, when Norman Triplett's study showed that people executing simple tasks performed better when others watched them: social evaluation influenced the performance (Guerin, 2009: 7). Famous in this respect is the study by Stanley Milgram of the willingness of people to obey authority, showing the ease with which ordinary people obey a 'legitimate' authority and appear to be willing to perform cruel acts against innocent victims (discussed in Blass, 2009). Thaler and Sunstein's *Nudge* focuses on this idea of social context as well, but they expanded the idea far beyond authority; among other influences they emphasise social pressure and the need to fit in.

Finally, emotion, mental shortcuts and intuition are explored in the work of Kahneman (2012), which is considered ground-breaking for the field of behavioural economics. He pointed out the heuristics people use, and the biases people are led by, in situations of uncertainty, such as loss aversion (see Anger and Loewenstein, 2006: 30). Kahneman developed Prospect Theory, which describes how people are primarily led by perceptions of losses and gains, rather than by the final outcome. It shows that the way options and choices are framed, i.e. as losses or as gains, is what is decisive for the choices people make (ibid., p. 31). If we accept behavioural economics, we accept that we are no longer driven solely by economic rationale, but also by our 'flawed' emotional drivers.

Thaler and Sunstein's theory of 'choice architecture' encompasses the characteristics of the environment or context in which a choice or decision is made, and the term was coined in their work *Nudge* (2008). Sharla Stewart, the author of 'Can behavioural economics save us from ourselves?', puts it like this (cited in Rainford and Tinkler, 2011: 4): 'Once you know that every design element has the potential to influence choice, then either you close your eyes and hope for the best, or you take what you know and design programs that are helpful.' Thaler and Sunstein propose to 'nudge' people – by means of better choice architecture – in the 'right' direction. The reason people go in the 'wrong' direction is understood to be because they often rely on their automatic systems. Thaler and Sunstein confirm that, on one level, we think intuitively and act automatically (the 'automatic system'), while on another level we think rationally and reflect before we act (the 'reflective system').

To perform an analysis of non-rational processes of decision-making, a useful starting point can be found in Thaler and Sunstein's work. They make a distinction between the biases people are susceptible to due to their tendency to rely on the automatic system and the types of nudge that might be used as possible solutions. In Table 4.1 we have made a brief summary of the biases, and in Table 4.2 we

Table 4.1 Biases

Biases	What do we do?
1 Points of reference	The information we already have influences how we judge other situations, even if our former knowledge and new knowledge are not connected in any way.
2 Availability	We suppose the possibility of something happening based on the availability of examples that come to mind. The more you hear something, the more often you think it happens.
3 Representativeness	We tend to construct patterns, confusing random fluctuations with causality.
4 Optimism and overconfidence	We overestimate our own capacities and are unrealistically optimistic about our chances.
5 Gains and losses	We hate losses disproportionately when compared to how much we like winning. Bluntly stated, losing something upsets us twice as much as gaining the same thing makes us happy.
6 Status quo bias	People have a general tendency to cling to their current situations.
7 Presentation	We are influenced by how options are presented to us: when, by whom, what words are used?
8 Temptations	People have difficulties fulfilling their own expectations. Besides this, unreflected or 'mindless' choices also have a large influence on people's day-to-day activities.
9 Following the herd	We are easily influenced by what other people say and do.

Table 4.2 Nudges

Nudge	What does it do?
1 Anchors	An anchor is a point of reference. By adapting an anchor to a different reference point, we are prone to judge situations differently.
2 Default option	A default option is that which happens if we do nothing. By changing the default option to a more desirable outcome for most people, a more desirable outcome will most likely be the result.
3 Feedback	If you give information about people's behaviour by means of direct feedback, including a comparison to social norms or to what the desired behaviour would be, this can be more effective.
4 Framing	We often accept information as it is handed to us. By framing information differently it will be perceived differently, i.e. as a loss, as a gain or as a continuum/stability.
5 Incentives	People may be stimulated by positive or negative incentives that not only influence rational calculation but also target psychological mechanisms, such as comparison to peers.

	Nudge	What does it do?
6	Priming	By making certain characteristics of an environment more prominent, it can remind people of certain goals or values.
7	RECAP: Record–Evaluate–Compare Alternative Prices	By giving insight into relevant details about the costs of people's behaviour, people can evaluate and reconsider their choices.
8	Required choosing	People may be obliged or stimulated to make a deliberate, conscious choice (a variation and softer version of the default option).
9	Reminders	When we remind people of the risks or values of a certain choice, they will be less likely to act based on their overconfidence or loss aversion.
10	Reversal of loss aversion	We may make use of loss aversion to motivate people to do something.
11	Structure complex choices	Making a choice simpler, by reducing options or by taking care of complex aspects of a decision, means people will find it easier to make a decision.

have summarised different modes which are deemed in this body of work to be capable of guiding non-rational behaviour – usually known as 'nudges'.[2]

Thaler and Sunstein use the (apparent) oxymoron of 'Libertarian Paternalism' to characterise the efforts of government to influence the choices people themselves make. There is debate about whether or not nudging is an intrusion on free will. Thaler and Sunstein claim libertarian paternalism is relatively 'nonintrusive' and essentially liberal, as it centres on people's own personal choices. The essence of nudging is that individual choice is free, but influenced and steered at the same time. Thaler and Sunstein argue that more deliberate design of choice architecture will help people to choose what they really want and therefore strengthens people's personal choice. For Thaler and Sunstein, nudging aims to 'maintain or increase freedom of choice' (Thaler and Sunstein, 2008: 5; see also van Oorschot et al., 2013).

The theories of libertarian paternalism and nudge have received much criticism, focusing on both practical and ethical issues. First of all the practical issues: Selinger and Whyte (2011, 2012) argue that nudges are not sufficiently demarcated, which is problematic. Thaler and Sunstein explain that certain situations call for nudges, while others do not. According to Thaler and Sunstein, 'choice architect friendly' situations are complex decisions that are made once so the decision-maker does not get much practice and decisions where the outcomes will have consequences in the future, so the costs and benefits are unevenly distributed over time; due to this uneven distribution there is no 'feedback-friendly' structure. Even though the terms are clear, they are ambiguous. This ambiguity is problematic since a nudge exists by virtue of transparency: when it is unclear whether a policy area is 'choice architect friendly', it is unclear whether a nudge should be applied.

Besides the discussion on the practicality of nudge, there is also an ethical discussion: should we want nudge? Some claim libertarian paternalism is a threat to individuals' control over their own decisions, a dangerous manipulative tool (Hausman and Welch, 2010; Goodwin, 2012). But, as Glaeser (2005) argues in reaction to Sunstein and Thaler's earlier publication on libertarian paternalism (2003), paternalism might be here to stay, so guiding 'soft' paternalism might be a key priority in that context. Johnson et al. (2012) support this view and explain that choice architecture is unavoidable and can have a positive effect on 'real-world decisions'. Meanwhile Maio et al. (2007) fear that using behaviourism, moral norms and peer pressure as tools for policy-making will create moral stigmas against people who do not live up to these norms. In line with this thought we might need an ethical framework in order to responsibly apply these nudging techniques (Selinger and White, 2011; Blumenthal-Barby and Burroughs, 2012). Finally, Leonard (2008) explains that libertarian paternalism has its inconsistencies. It dismisses the *homo economicus* as a realistic view of humans but embraces that same image as a final goal: we should be nudged towards economic rationality.

In an attempt to counter most of the ethical criticisms, Thaler, Sunstein and Balz (2010) explain that:

> Decision makers do not make choices in a vacuum. They make them in an environment where many features, noticed and unnoticed, can influence their decision. The person who creates that environment is, in our terminology, a choice architect.
>
> (abstract)

They argue every decision is made within a constructed framework, this framework influences the way people decide, and adaptations of the framework will affect the final decision. For this reason alone, it is relevant to research how these frameworks affect decisions and how these frameworks may be adjusted. That is exactly our challenge in the next section: exploring how the design of the Dutch housing market affects decisions by buyers and sellers and how this design may be adjusted to stop the 'lock-in'.

From theory to policy advice

In this section we attempt to show what steps might be required in order to turn the aforementioned theories into realistic tools for policy-makers. As a policy-maker, what nudges can you use, and in what situations? This section is largely based on the exploratory advisory work the Netherlands School of Public Administration has done for two Dutch ministries, the Ministry of Internal Affairs and the Ministry of Infrastructure and the Environment. The question was to further explore lock-in on the Dutch housing market (e.g. Struyven, 2015; Høj,

2011) with a focus on the possibility of non-rational interventions to circumvent said lock-in.

The lock-in of the housing market in the Netherlands

The financial and economic crisis that has challenged the world since 2008 has had a significant impact on the housing markets of many Western countries. The Netherlands is no exception. The housing market in the Netherlands has several specific characteristics that have even worsened the crisis. For example, there is relatively high income tax deduction on mortgage interest payments, which compensates the costs of interest for house-owners. For medium and higher incomes this means that the net interest of a mortgage is cut in half. This has been so for decades and has resulted in the co-existence of high mortgages alongside relatively high interest rates. People only 'feel' half of the interest they pay; the rest is covered by public subsidies. Over the years, the stream of cheap credit (mortgages) spurred the demand in the housing market and pushed prices up, leading to higher prices and even higher mortgages. Also, many people have 'interest-only mortgages' because the large subsidy on interest disincentivises payoff of the mortgage. However, when housing prices dropped in the early phase of the financial crisis, many house-owners were pushed into negative equity; they have a mortgage that exceeds the value of their house. Apart from problems with renewing their mortgage with the bank, this leads to huge problems when selling a house. Many are literally locked into their house and unable to move. The demand for houses has spiralled downwards into a self-fulfilling cycle of lower prices, inability to sell at the new market price, fewer transactions, even less demand, even lower prices, more pressure from banks on mortgage loans, even less demand: the housing market in the Netherlands is locked into a vicious circle.

Biases in the Dutch housing market

In response to the fact that the housing market in the Netherlands, as in many other countries, had come almost to a stop because of the financial and economic crisis that started in 2008, the Dutch Ministry of Housing asked our research team to help them rethink the role of 'non-rational' behaviour to reintroduce dynamism into the housing market, thereby acknowledging 'all kinds of emotions' (Christie et al., 2008: 2297) that guide the behaviour of buyers and sellers.

In order to analyse the vicious circles in which the Dutch housing market was locked, we used a simple conceptual model, consisting of the biases and nudges presented in Tables 4.1 and 4.2. We organised workshops and reflection sessions with officials from the Ministry of Housing. These sessions were aimed at: (1) sharing insights from academic literature and discussing their relevance for policy practice; (2) developing shared participation information and nudge options in rethinking the role of 'non-rational' behaviour to reintroduce dynamics in the housing market, and (3) developing and reflecting on preliminary views, beliefs and analyses. We gathered project documents and internal correspondence. In addition, we performed a content analysis of relevant policy documents. This was done to strengthen our understanding of the developments in the policy field. We also had frequent face-to-face and e-mail conversations with several civil servants from the ministry over the course of the project and did interviews with other stakeholders around and during participation meetings. To facilitate ongoing reflection, we used memos to document preliminary observations and interpretations over the course of the project, shared preliminary analyses with a peer group and asked the ministry to comment on a preliminary version of our research report.

From this analysis we found several non-rational biases (under some of the headings in Tables 4.1 and 4.2) that may be responsible for the persistence of the lock-in of the Dutch housing market. These are:

- *Points of reference*: When deliberating on a change in their housing situation, homeowners attempt to estimate the value of their home in order to make a financially wise decision. Often, however, homeowners estimate the value of their home solely on the basis of the price increase on the market, instead of taking into account *all* the points of reference, which include many more factors, for example the prices of other houses, inflation, the overall situation of the financial market, etc. Moreover, the point of reference used should be (but rarely is) the real housing costs (rent or mortgage, but also interest, inflation, down payment, etc.) instead of purely the initial investment.
- *Optimism and overconfidence*: Since people believe that the value of their house should rise and not decline, there is no reason to sell their house if they are selling at a loss. Before the housing market fell, property was seen as a sound investment, as a forever increasing investment. As a result, people are still unrealistically optimistic about the market and the value of their own house.
- *Framing*: For a long time the Dutch government tried to stimulate individuals towards buying: paying rent was portrayed as throwing money away. If you have a house, you are investing in yourself. If you are paying rent, you are investing in your landlord. The ideal was to have your own house. As a consequence, people took on mortgages that they could only just afford in order to realise the ultimate ambition of home ownership. With decreasing prices and the tightening of tax deduction rules, this ambition has become a burden.
- *Following the herd*: We performed a media analysis on Dutch national newspapers from 2008 to 2012 (available in Dutch upon request). This found the dominant view was that the housing market is 'locked'; it will not get better soon, no extra investments will be made by the government, and all players are guilty. If the media reports that all are holding on to their current properties/housing, people tend to follow that trend. The housing market is "locked" because people see and hear that it is, and they consequently act as if it were locked. The locked-in discourse becomes a self-fulfilling prophecy.

Potential nudges in the Dutch housing market

Building upon this analysis, we formulated a set of non-rational interventions based on the principle of nudge that may be helpful in releasing the lock on the housing market. These were:

- *Anchors*: It is complicated to compare a mortgage with rent. If a single anchor for *both* types of payments were used, a good anchor, a good reference point for both would make a decision easier. The way forward here would probably be to develop a common standard for the costs of living, including all costs that should be included in calculating the relative costs of home-ownership and rental. This would also include the costs for write-offs that homeowners tend to overlook.
- *Framing*: Up until now, buying a house has been framed in policy as ideal and renting as a way of throwing away money. When people think of home-ownership, they think of security and stability. When people think of renting, they do not necessarily see a utility that costs a monthly sum and compare it with costs of ownership since they see a house as an investment that should pay off in the long run. Therefore, people tend to stay in a small rental property while saving money to buy, whereas in their situation it may be beneficial to move to a bigger rental before buying. By framing the current options in the housing market differently, by easing away from the idea that homeownership by definition is the safer choice, the problems mentioned could be bypassed.
- *Reversal of loss aversion*: If we are aware that people shy away from certain actions due to loss aversion, it explains why people refuse to sell their house even when their mortgage is an unmanageable financial burden – the idea that they will be losing money weighs heavier than the reality that the situation will only become less manageable. An alternative here is telling people not that they must minimise their losses and sell now, but that by selling now they will have more money to invest in another house in future. The way in which the information is presented is different, not the bottom line of the message, which is: you should sell your house. Another way to counter loss aversion may be simply to accept the fact that people shy away from loss. By minimis-ing this loss (through tax breaks or by partially remitting [former] homeowners from their residual debt) the effects of this bias could be reduced.
- *Structure complex choices*: The housing market is extremely complex, and the numerous layers of making a decision concerning housing (price, neigh-bourhood, future trends, etc.) are equally complex. By dissecting these layers, and structuring complex choices, making a decision could become easier.

There is much debate about possible interventions in the housing market. Some argue that more financial subsidies could strengthen the demand for houses. From a behavioural perspective a different set of options becomes more feasible. The market is not locked because it is, but because people act as if it is. The lock-in is not an objective barrier to entering the market, or an act of nature, but is the combination of different behaviours of individual agents on the housing market and wider structures and framings, such as those of government tax policies and banking behaviours. Paradoxically, the destructive self-fulfilling downward spiral of the housing market can be challenged and changed only by different behaviour by the same consumers that are caught in the downward spiral. If people accepted an often largely subjective loss on their house and started to see a house as a utility that costs a monthly sum rather than as a solid investment that generates money in the long term, some of their choices would change. Once such dynamics started, a reversed cycle could set in; increased demand leads to higher prices, more transactions, more demand, higher prices, renewed confidence, even more demand, and so on.

Knowing this, we accept that people make decisions based not only on their rationality, but also on their emotions. Consequently we suggested in our workshops that policy-makers should use tools to guide citizens' behaviour that account for this emotional motivation alongside the rational motivation. Nudge does just this: by connecting the policy tool to the biases in question, a nudge is created.

Practical issues and moral dilemmas of nudging

In the previous section we showed how nudging might contribute to the 'unlocking' of the Dutch housing market. We also presented criticism of Sunstein and Thaler's ideas. From the Dutch housing case we can now illustrate some of the practical and moral dilemmas that are involved with the potential application of non-rational and emotional interventions in policy. We found that while most policy-makers were intrigued by the idea of nudging and a more non-rational approach to policy-making, there was also reluctance to change the way in which policy was made: bureaucracy limits flexibility, and flexibility is necessary for nudge to be applied as long as it is not a standard policy-making tool. We found that on a small, yet widely spread, scale, policy-makers were informing themselves on nudge and were exploring in what manner nudge could be applied. The non-intrusive nudges (such as framing) seemed more popular than more intrusive nudges (such as changing a default option) – this is most likely caused by how nudge is organised in the Dutch government, which brings us to the practical dilemmas of nudging.

Practical dilemmas of nudging

First, a very practical and general dilemma is how to position nudging in the organisation of policy. There are numerous countries that have set up special units in executive offices of government, such as the Behavioural Insights Team (BIT) in the USA and UK (Whitehead et al., 2014: 23; see also Pykett et al., Chapter 5 in this volume). These units are positioned outside the normal flow of policy and work, and step in on dossiers/issues selected by them. An alternative would be not to organise a separate unit but to integrate nudging into the regular modes of policy-making, for instance by integrating it into the toolkit and checklists that policy-makers apply when making policy or designing implementation schemes. This is how the OIRA functions, the Office of Information and Regulatory Affairs of the USA; situated in the executive office, OIRA reviews all drafts of proposed and final regulations. The practical dilemma is essentially a question of whether to integrate the professional knowledge and ethical balancing acts on nudging into the mainstream of policy-making or to keep them separate, and whether or not nudge should be applied throughout the entire government (as is the case in the US and UK) or only in demarcated policy fields (as is the case in the Netherlands where nudge is applied very specifically within the Ministry of Infrastructure and Environment).

Second, a practical dilemma remains as to whether nudge as a policy tool should remain an experimental, non-standardised tool or should be institutionalised, possibly limiting its political impact. For existing policy methods there are literally decades of experience in demarcating the domain of political democratic decision and bureaucratic competency. The procedures and norms are clear and known, and the checks and balances are public. However, in the case of nudges the rules and procedures regarding moral and ethical deliberations are still, at best, in an early stage of development. The working processes and organisational protocols have not solidified yet and in most cases are entirely absent. Therefore, nudging remains a 'free space' with a lot of unaccountable room for bureaucratic experimentation with potentially large political impacts (both positive and negative).

Moral dilemmas of nudging

In addition to the practical dilemmas, there are also moral dilemmas related to the practice of nudging. The legitimacy of nudge is often debated. When we reflect on our analysis of the Dutch housing market, the following moral dilemmas emerged.

First, what role should the government have in individual decisions? This is an everlasting discussion on government intervention, but one seen through a new lens since citizens are supposedly not aware they are being nudged. This is the classical division between libertarians and statists, wherein libertarians see nudge as a 'repackaging of big government' (Leggett, 2014: 7), and statists claim that libertarian paternalism is pro-market and anti-state. This raises questions about whether good and bad life decisions should be up to the individual, whether government should impose choices in a manner that is not visible for the citizens, what rules policy-makers should adopt, and whether libertarian paternalism is truly as non-intrusive as Sunstein and Thaler state or is instead a reframing of paternalism.

Second, the theory of nudge assumes that a rational decision is clouded by an individual's automatic system. The theory of nudge also assumes that ideally all individuals choose on the sole basis of rational deliberations. However, if we accept that individuals are not fully rational, should we not also accept that their decisions are also not fully rational? This dilemma touches upon the core of the critique of nudge as it questions the intentions of nudge (nudging people towards rationality).

Third, there is a trade-off between the effectiveness and the openness of nudging, raising issues of how honest and open government should be about nudging. In policy proposals it is common to clearly explain the actions that are to be undertaken and what the policy theory behind them is. For nudging, it is probably more difficult to be open and transparent; at least some nudges probably greatly lose in effectiveness if the targeted audience knows about how they work. This forms part of a larger debate in which the deliberative democratic approach is opposed to that of choice architecture (John et al., 2009: 361). Following our analysis, however, we were convinced that, by offering insight to the public biases and the connected nudges, a similar effect could be reached to operating in a more hidden manner. The public might be helped more with a deconstruction of choice architecture than with an adapted choice architecture.

Conclusions

This chapter has explored the potential of a non-rational, emotional perspective on a thorny policy problem: the case of the Dutch housing market. We have seen that nudges may offer some new and innovative opportunities for interventions: they stretch the possibilities that traditional policy tools offer by accounting for the biases that influence people's decisions. In the Dutch housing market this could, for example, mean that homeowners and renters are no longer encouraged to believe that homeownership is to be preferred to renting, allowing for both to assess their own situation freely, with as few prejudices as possible. The market could be simplified by structuring complex choices and by using a singular way of presenting monthly costs (instead of dealing with mortgages versus rent); those in the housing market might create an environment in which actors are as informed as they can be before they make a decision.

However, even though nudges offer practical opportunities, there are also practical worries. Who will be responsible for these nudges? How can a nudge be transparent? What rules should be followed when using nudges in policy? The moral questions that arise in respect of nudge, mostly grounded in the need to protect the freedom of each individual and thus rejecting any form of manipulation, cannot be ignored. When looking for new solutions in the emotional governance of nudging, we must inevitably also engage in a new discussion on what practical and moral issues arise when we change the nature of public policy-making. Future research is required to examine how policy-makers deliberate on why they do or do not apply nudge, and what obstacles they experience when converting the theory of nudge in the policy context into policy reality.

Notes

1 In this year, a behaviourist manifesto was published by J.B. Watson: 'Psychology as a behaviourist views it' (see Jones and Elcock, 2001: 51).
2 These tables are primarily based on the work of Kahneman (2012) and Sunstein and Thaler (2008). In addition, we have identified 'nudges' that have already been proposed or put into practice by the Behavioural Insights Team of the Cabinet Office in the UK

(COBIT 2010, 2011a, 2011b, 2011c, 2012a, 2012b; Cabinet Office, 2011) and the Office of Information and Regulatory Affairs (OIRA) in the USA (Analysis of Review Counts by OIRA from 2002 to 2012, data available at: www.reginfo.gov/public/do/ eoCountsSearchInit?action=init (accessed March 2013).

References

Anger, E. and Loewenstein, G. (2006) 'Behavioural economics', in Mäki, U. (ed.) *Handbook of the philosophy of science*, Vol. 5, at: www.cmu.edu/dietrich/sds/docs/ loewenstein/BehavioralEconomics.pdf (accessed 13 March 2013).

Blass, T. (2009) *Obedience to authority: current perspectives on the Milgram Paradigm.* Psychology Press, London and New York.

Blumenthal-Barby, J.S. and Burroughs, J. (2012) 'Seeking better health care outcomes: the ethics of using the "nudge"', *American Journal of Bioethics*, 12(2): 1–10.

Bovens, M.A.P., 't Hart, P. and Van Twist, M.J.W. (2012) *Openbaar Bestuur.* Kluwer, Deventer.

Cabinet Office (2011) *Government response to the Science and Technology Select Committee Report on Behaviour Change*. Cabinet Office, London.

Christie, H., Smith, S.J. and Munro, M. (2008) 'The emotional economy of housing', *Environment and Planning A*, 40: 2296–312.

COBIT (2010) *Applying behavioural insight to health*. Cabinet Office, London.

COBIT (2011a) *Better choices, better deals: consumers powering growth.* Cabinet Office, London.

COBIT (2011b) *Behavioural change and energy use*. Cabinet Office, London.

COBIT (2011c) *Annual update 2010–2011*. Cabinet Office, London.

COBIT (2012a) *Applying behavioural insights to reduce fraud, error and debt*. Cabinet Office, London.

COBIT (2012b) *Annual update 2011–2012*. Cabinet Office, London.

Glaeser, E.L. (2005) 'Paternalism and psychology', *Harvard Institute of Economic Research, working paper 2097*. Cambridge, MA.

Goodwin, T. (2012) 'Why we should reject "nudge"', *Politics*, 32(2): 85–92.

Guerin, B. (2009) *Social facilitation*. Cambridge University Press, Cambridge.

Hausman, D.M. and Welch, B. (2010) 'Debate: to nudge or not to nudge', *Journal of Political Philosophy*, 18(1): 123–36.

Høj, J. (2011) 'Improving the flexibility of the Dutch housing market to enhance labour mobility', *OECD Economics Department working papers*, No. 833. OECD Publishing, Paris.

Hothersall, D. (1995) *History of psychology.* McGraw Hill, Columbus, OH.

John, P., Smith, G. and Stoker G. (2009) 'Nudge nudge, think think: two strategies for changing civic behaviour', *Political Quarterly*, 80(3): 361–70.

Johnson, E.J., Shu, S.B., Dellaert, B.G.C., Fox, C., Goldstein, D.G., Häubl, G., Larrick, R.P., Payne, J.W., Peters, E., Schkade, D., Wansink, B. and Weber, E.U. (2012) 'Beyond nudges: tools of choice architecture', *Marketing Letters*, 23(2): 487–504.

Jones, D. and Elcock, J. (2001) *History and theories of social psychology: a critical perspective*. Routledge, New York.

Kahneman, D. (2012) *Thinking fast and slow*. Penguin, London and New York.

Leggett, W. (2014) 'The politics of behaviour change: nudge, neoliberalism and the state', *Policy and Politics*, 42(1): 3–19.

Leonard, T.C. (2008) 'Review of Richard Thaler and Cass Sunstein, Nudge: improving decisions about health, wealth and happiness', *Constitutional Political Economy*, 19(4): 356–60.

Maio, G.R., Verplanken, B., Manstead, A.S.R., Stroebe, W., Abraham, C., Sheeran, P. and Conner, M. (2007) 'Social psychological factors in lifestyle change and their relevance to policy', *Social Issues and Policy Review*, 1(1): 99–137.

Rainford, P. and Tinkler, J. (2011) *Designing for nudge effects: how behaviour management can ease public sector problems.* Originally presented at Designing for Nudge Effects: How Behaviour Management Can Ease Public Sector Problems, Innovating through Design in Public Sector Services seminar series, 23 February 2011, LSE Public Policy Group, London.

Selinger, E. and Whyte, K.P. (2011) 'Is there a right way to nudge? The practice and ethics of choice architecture', *Sociology Compass*, 5(10): 923–35.

Selinger, E. and Whyte, K.P. (2012) 'What counts as a nudge?', *American Journal of Bioethics*, 12(2): 11–31.

Struyven, D. (2015) 'Housing lock: Dutch evidence of the impact of negative home equity on household mobility', *MIT Job Market Paper*, 12 January 2015.

Sunstein, C.R. and Thaler, R.H. (2003) 'Libertarian paternalism is not an oxymoron', *University of Chicago Law Review*, 70: 1159–203.

Thaler, R.H. and Sunstein, C.R. (2008) *Nudge: improving decisions about health, wealth and happiness*. Yale University Press, London.

Thaler, R.H., Sunstein, C.R. and Balz, J.P. (2010) *Choice architecture*. At SSRN: http://ssrn.com/abstract=1583509 (28 June 2016).

Van Oorschot, K.A.H., Haverkamp, B., Steen, M. van der, and Twist, M. van (2013) *Choice architecture*. Working paper NSOB, at: www.nsob.nl/wp-content/uploads/NSOB_Choice-Architecture-web.pdf (accessed 9 December 2014).

Whitehead, M., Jones, R., Howell, R., Lilley, R. and Pykett, J. (2014) *Nudging all over the world*. Economic and Social Research Council, at: https://changingbehaviours.files.wordpress.com/2014/09/nudgedesignfinal-1.pdf (accessed 25 October 2014).

5 Governing mindfully

Shaping policy makers' emotional engagements with behaviour change

Jessica Pykett, Rachel Howell,
Rachel Lilley, Rhys Jones and
Mark Whitehead

Introduction

'Behaviour change' has become something of a policy panacea across a range of social policy sectors worldwide. There is of course nothing new about the shaping of citizenly conduct. Sophisticated tools of persuasion and more blunt tools of compulsion have long been deployed by state authorities and non-state actors alike. But since at least the mid-2000s, concerted efforts have been made by several national governments to better understand the psychological parameters of decision making contexts and ingrained human biases. Van Oorschot and colleagues' discussion of the use of nudges in Dutch housing regulation in the preceding chapter (Chapter 4) is one case in point. These efforts have been informed by research findings from the disciplines of behavioural economics, social psychology, marketing, design and neuroeconomics – as well as by the popularisation of such disciplines by books such as *Nudge* (Thaler and Sunstein, 2008). As a result, the 'behaviour change agenda' (Jones et al., 2013) has emerged as a catch-all term for adopting a behavioural science approach to public policy. Central to this agenda is a conceptualisation of human decision making simultaneously as a cognitive and non-cognitive process; our decisions are not limited to what is going on in our conscious and reflective minds, but are informed as much by our affective 'moods' and immediate contexts. 'Changing behaviour' relies on the development of new forms of emotional and psychological governance which target both the situations in which decisions are made and the emotional drivers of those decisions.

The behaviour change agenda also changes the nature of policy making insofar as the forms of expertise and insight required in some senses are claimed to 're-humanise' the policy making process. This means that policy makers themselves are required to collect behavioural insight on their target audiences/citizens, pay more attention to how policies will be implemented in light of apparently improved understandings of the human condition, and design policies which are emotionally informed (e.g. Dolan et al., 2010). This chapter investigates how policy makers are engaging with behaviour change knowledge, including how this engagement resonates with the emotional demands already placed upon them within their working lives in the public sector. In doing so, it considers this increasing influence of behavioural insights on the policy making process alongside the contemporary focus on workplace wellbeing, which is one sector in which theories from behavioural science have been mobilised to change workers' behaviour and organisational cultures. Thus civil servants in the UK context and elsewhere are coming into contact directly and indirectly with both of these current trends.

The chapter focuses on participatory action research which the authors undertook with a group of Welsh Government civil servants in Cardiff and Aberystwyth in 2014. This was part of a larger research project on 'Negotiating Neuroliberalism'[1] within policy contexts, which has examined how the human subject is being reconceptualised as vulnerable to cognitive biases, mental shortcuts and irrationality – and thus amenable to a wide range of hitherto untested behaviour change techniques. This wider research has to date involved a desk-based mapping exercise to identify the international spread of public policy initiatives informed by behavioural science, international case studies of the emergence of behaviour change polices in Australia, the Netherlands, Singapore and the USA, and semi-structured interviews with behaviour change practitioners in the UK from within public policy, the third sector and commercial organisations.

The strand of research reported on here sought to develop a more participatory form of research intervention as a useful way to open up the behaviour change agenda to further critical scrutiny. Having felt that we had to a certain extent exhausted our capacity to offer critique of the behaviour change agenda (as provided for instance in Jones et al., 2011, 2013; Whitehead et al., 2011, 2012; Pykett, 2012), we wanted to develop a socially impactful research study which retained a broadly questioning stance towards the venture of behaviourally informed public policy. Our aim was to do so in a way which *both* brings out important limitations of existing narrow accounts of evidence-based policy evaluation (often informed solely by behavioural science disciplines) *and* provides a route forward for developing (potentially) more progressive forms of emotional governance. This stance reflected the varied commitments of the research team towards engaging in critical theoretical analysis to challenge key assumptions of the behaviour change agenda, examining the positive potential for behaviour change (primarily aimed at securing an environmentally sustainable future, though this is not discussed directly here), and developing practical behaviour change interventions in the third sector with the broad social purpose of community development and regeneration. A certain internal conflict is therefore manifest in this chapter as we try to reflect on the possibility for negotiating these different and potentially competing aims.

From April 2014 to October 2015, the authors developed and delivered a 'Mindfulness, Behaviour Change and Engagement in Public Policy' (MBCEPP) programme in partnership with the Welsh Government, who were interested in helping staff to better understand theories of behaviour change. The programme sought to investigate the potential value of mindfulness practice as a means for participants to engage more critically with insights about the nature of human behaviour and how it can be changed, as relevant to their diverse working remits (particularly behavioural biases, rationality, habit formation, emotional drivers of decision making, heuristics). The aim was to trial a new method for civil servants to engage with behavioural insights in a way which specifically addressed a number of critiques and sought to be empowering, effective and ethical. It was explicitly *not* intended to be a form of therapy or stress relief. Rather, it was intended to bring to participants' attention the potential ethical dilemmas associated with subconscious 'nudge' techniques intended to guide – but not force – people to make decisions in their own best interests, by informing the civil servants about specific behavioural insights highlighted in behavioural economic research. Previous research has found mindfulness meditation to be effective in 'de-biasing' common cognitive heuristics, such as the tendency to perpetuate 'sunk costs' in economic decision making in particular (Hafenbrack et al., 2013). The MBCEPP course was designed not simply to teach policy makers about behaviour change approaches (the majority already had prior knowledge of these). Rather, it was designed to address some of the existing critiques of behavioural forms of emotional governance in public policy, including our own which are explored below. Our formulation of mindfulness practice was *both* concerned with an embodied awareness (something central to mindfulness practice – see below) *and* discussion and collective reflection on how the techniques of mindfulness might enable participants to learn about their minds, behaviours and contexts and those of the 'recipients' of policy. The programme was therefore designed to go beyond present experiences of civil service practice and encourage reflexive 'dialogue with the self' and others (in this case work colleagues), in the fashion described in previous geographical scholarship on mindfulness and emotional habits by Lea et al. (2014: 15; see also Bissell, 2011). The chapter advances this work by directly addressing the potential political ramifications of mindfulness-based learning about behaviour change within the public sector. We wanted to investigate how what we argue to be a distinctly geographical formulation of mindfulness practice could be developed – a form of practice which emphasises not just individual wellbeing but addresses context-specific issues of social injustice, collective suffering and political struggle, aspects which are arguably inherent in the roots of mindfulness (Kabat-Zinn, 2005).

Our research team included a mindfulness teacher and researcher, who delivered a taster workshop and an eight-week programme involving three sets of civil servants, over 18 months, a weekly two-hour workshop reaching 40 civil servants, in addition to support between the sessions via resources and emails. The participants also engaged in 'everyday' personal mindfulness practice (recommended 4–5 times per week 10–20 minutes and short practices during the day of 1–2 minutes). The course leader followed nationally established good practice for mindfulness teachers.[2] We evaluated the programme through a pre- and post-intervention survey – one section determining participants' prior knowledge of the principles of behaviour change and the other section based on the 39-point Five Facets of Mindfulness Questionnaire (Baer et al., 2006), qualitative interviews with participants, and a feedback/feed-forward workshop facilitated by a senior civil servant responsible for change management at the Welsh Government.

The chapter proceeds as follows. In the next section ('An apple a day'), we introduce some of the main precepts of behaviourally inflected public policies in the UK and outline the global significance of this type of approach. We summarise critical perspectives on the emergence of 'behavioural governance' as a prelude to our discussion of the progressive potential of the MBCEPP programme later in the chapter. The section on 'a second apple' considers the manifestation of behaviour change in workplace wellbeing programmes, an area hitherto unexplored within human geography. The subsequent section introduces the experience of delivering the MBCEPP programme in order to show how civil servants can engage emotionally and in embodied ways with behaviour change knowledge in a manner which is potentially empowering, effective and ethical. This recognises that policy makers are themselves subject to the same cognitive biases as those attributed to the 'post-rational' citizen. The conclusion then reflects on some of the limitations of this kind of participatory research, and in particular the political tensions inherent in research which takes the stance of positive critique.

An apple a day: mainstreaming global behaviour change

Policies influenced by behavioural insights are based on an account of the human mind as fallible, susceptible and systematically irrational, although staff at the Behavioural Insights Team, set up in the UK in 2010 to develop behavioural science informed policy making, have begun to express their reservations about the appropriateness of the terms 'irrational' and 'biases' to express the concepts of human subjectivity found in behavioural science literatures (Hallsworth and Sanders, forthcoming). Nonetheless, this new vision of personhood is in stark contrast to that of rational economic man forwarded by the abstract economic models which have hitherto dominated policy making. Behavioural scientists argue that because we tend to be 'loss averse', straightforward incentive schemes may not always be the best way to change behaviour because people respond more strongly to the threat of losing something than gaining. Furthermore, since we tend to 'discount the future' in our decisions, it is argued that we need policies which make it easier to plan ahead, for instance by automatically enrolling us in pension schemes. Because we 'anchor' our decisions to the most salient pieces of information that we have around us, our decisions can be 'primed' or swayed by what we have just been told or by our immediate environment – for instance we only repay very small amounts on our credit cards because the minimum payment is set incredibly low. Or we might drink unhealthy amounts of alcohol because our wine glasses are too large (Dolan et al., 2010: 24, 44).

Understanding the emotional drivers of decision making is thus regarded as central to the successful application of behavioural insights. Policy strategists promoting the behaviour change agenda tend to take a 'two-brained' model of human decision making (see Schüll and Zaloom, 2011, for a full account). Behavioural economist Daniel Kahneman refers to 'thinking fast and slow' (2011: 20–1) to distinguish between the brain's two processing systems. First is the automatic system, which is unconscious, easy, fast and emotional. It relies on impressions, intuition and feelings, and is easily triggered into action. Second, there is a reflective system which is more deliberate, requires effort and concentration, is slow and rational – and, by implication, more intransigent to change. Public policy informed by behavioural insights therefore appears, on the one hand, to be somewhat mistrustful of the emotions, seeking to direct citizens away from hasty, intuitive judgements made in the heat of the moment. On the other hand, attempts to govern explicitly through prompting certain emotional responses can be important strategies for behaviourally informed governance, for instance through direct targeting (or explicit overriding) of affective responses, through capitalising on social relationships and trusted messengers, or through harnessing an expressed and felt commitment to change (Dolan et al., 2010).

Governing 'the *MINDSPACE* way'[3] – a shorthand for describing how behavioural insights should inform policy making (Dolan et al., 2010) – places attention on the minds of citizens and the immediate spaces in which they make decisions, heralding the emotional citizen as a new subject of psychological governance, and the emotionally literate state as skilled in shaping decisions and 'choice architectures' (the contextual frameworks for decision making) (Thaler and Sunstein, 2008) for the public good. Public policy in the UK has been influenced by behavioural research for more than a decade across a range of sectors, from public health promotion and personal finance to pro-environmental behaviours. Behavioural experts have been embedded in UK government departments for some time, and periodic reviews and toolkits on behaviour change have been published for policy audiences in many different sectors (e.g. National Institute for Health and Clinical Excellence, 2007; Darnton, 2008; Chatterton, 2011). In 2014 the Behavioural Insights Team (BIT) was part sold as a social purpose company aimed at influencing governments and supranational policy forums around the globe. Our research on the global influence of behavioural insights on public policy has found, for instance, that the BIT has directly informed the establishment of centrally embedded policy units in Singapore, Australia, the Netherlands, Denmark and the USA, whist their work has been cited by the European Commission, World Economic Forum, World Bank and OECD (Whitehead et al., 2014). The BIT can therefore claim much success in creating the necessary conditions for establishing a successful central government unit to roll out behavioural insights across public policy. In a field replete with mnemonics,[4] the BIT offer APPLES (Administration, Politics, People, Location, Experimentation, Scholarship) as a 'juicy' prompt to help other national governments (presumably now their paying clients) to find the best way to mainstream behaviour change as a core public policy mechanism for ensuring the public good (BIT, 2015).

This global spread of government agencies with direct responsibility for behavioural insight and change raises pertinent ethical and political questions for critical social science researchers. How does the political, cultural, economic and social welfare context shape the delivery of behaviour change in different national contexts? How do different national publics respond to attempts to shape citizens' behaviour in ways which are said to be both freedom-enhancing and paternalistic? And why might the broader convergence between the corporate/

marketing techniques of a global behaviour change industry and state experts in behavioural science be significant? Several existing critiques have focused on the UK as a test bed for nudge-type policies. Peter John and colleagues (2011) have argued, for instance, that nudges tend to conflate freedom and choice, and that more deliberative and social forums for decision making are necessary to empower citizens to think, resolve conflict and make more rational collective decisions for the public good. Le Grand and New (2015) meanwhile have considered that (in light of the accumulation of scientific evidence on the failures of and limitations to citizens' decision making capacities) government paternalism should be justified only with respect to the mechanisms of public policies, and not their end goals. Leggett (2014: 10) takes the measured view that 'nudge' need not be incompatible with democratic ideals. Outlining the value of a 'reconstructive' social democratic 'behaviour change state', he outlines two key requirements. First, the state must be considered more than merely one agency of behavioural influence amongst many competing for our attentions by mimicking the private sector (Leggett, 2014: 4). As an arbiter of democratic preferences it must take an educative role in exposing the many nudges to which we are already subject as citizens. Second, there needs to be fuller understanding of the role of the 'social' in shaping decisions and norms. This understanding should highlight the ideological, material and historical production of such norms, including how these are structurally uneven, and should thus work alongside (rather than instead of) progressive social and economic policy.

In this chapter we want to argue that these kinds of largely sympathetic critiques are constructive on many counts, but we also want to go beyond them. Foremost, they chart a route forward for the pursuit of more transparent, empowering and participatory forms of behaviour change than might be implied by the initial policy enthusiasm for subconscious nudges. They foreground the value of knowledge about the world (not just about the self and behaviour) and highlight the political stakes of explicit efforts to direct the mind's attentions. However, we also want to acknowledge more staunch critics who have highlighted the alignment of the behaviour change agenda with market-based state strategies which prioritise a small state and reductions in public spending, with potentially disastrous consequences for already economically marginalised citizens (Wells, 2010; Cromby and Willis, 2014; Friedli and Stearn, 2015). Elsewhere, we have charted this alignment by investigating the co-evolution of neoliberal economic theory with the discipline of behavioural economics which has so influenced behaviour change in the international field of public policy making (Jones et al., 2013). Whilst not equating neoliberalism with the libertarian paternalist political philosophy underpinning nudges in particular, we have outlined the deployment of psychological and behavioural knowledges, cadres of experts and biomedical methodologies in ways which effectively psychologise a distinctively 'neuroliberal' state and society (Whitehead et al., 2011: 489; see also Rose, 1989). Foucauldian theories of governmentality and subjectification as well as his research device of genealogy have influenced our work to provide a critical account which serves to denaturalise the apparent inevitability and common sense of the behaviour change agenda. Specifically, this critical framework (Jones et al., 2013) interrogates how global social phenomena, e.g. climate change, obesity, financial crisis, are reframed as behavioural problems. We have therefore set out a way to think about how behavioural science knowledge as well as scientific research on the emotions have been used as forms of expertise by state authorities to forge both relatively small governmental initiatives and whole new programmes of government. Here we want to take this critique further – by showing what a practical behaviour change intervention informed by such a critique looks like and thereby to consider the political tensions involved in taking this research forward.

In sum, new forms of behavioural governance seen as being justified by several decades of behavioural science research, particularly in behavioural economics, have established novel governmental responses to the apparent problem that citizens do not always act in rational ways which would serve their own interests. Given the ethical and political questions raised above, it is vital to identify and understand how public policy makers are engaging – as emotional subjects themselves – with behavioural science knowledge in their working lives. At the same time, behavioural insights are being increasingly applied within human resources, organisational management and workplace wellbeing practices, including in public sector work, and it is to these applications that we now turn in order to identify the way in which the embodied and emotional subjectivity of the worker is implicated in practices of behavioural governance.

A second apple: positive minds at work

The 'second apple' of this section refers to efforts to improve workplace wellbeing. One such example is Mindapples, a UK-based social enterprise organisation which has been working since 2008 to 'help people understand how their minds work in order to improve their resilience and mental performance'. The basic premise of Mindapples is to support people to undertake simple habits and activities, "helping people to make the most of their minds, and about encouraging healthy, smart and sustainable working practices" (Andy Gibson, CEO of Mindapples, interviewed 5 March 2013). Just as eating five pieces of fruit and vegetables a day can be good for their bodies, the campaigning work of Mindapples appeals to people to do at least five good things for their mind daily, and to share these with others. They have so far amassed the suggestions and habits of over 100,000 people. This vast database, along with "bite-size insights from experimental psychology and neuroscience about how our minds work and how to use them",[5] informs their workplace training programmes and consultancy, which has been taken up by several large national and global organisations including JP Morgan, PriceWaterhouseCoopers, L'Oreal, the BBC and BUPA.

Workplace wellbeing training is therefore arguably a core locus for another kind of behaviour change. Here the term refers to work-based training initiatives diversely influenced by the science of positive psychology, affective neuroscience, psychometric and strengths-based assessments, management theory and secular writings on mindfulness. These might take the form of worker training programmes, change management and organisational schema, and discrete workshops, often delivered by outside consultants. Some such programmes are focused on stress relief and mental wellbeing. Others are aimed at improving productivity, personal resourcefulness, psychological resilience, leadership and workplace cultures. Many are a mixture of these things, such as the Mindapples example outlined above. All are concerned with improving the emotional positivity, wellbeing or happiness of workers and workplaces, and they can now be found across economic sectors within Europe, North America, Australasia, Asia and the Middle East, but particularly so in the knowledge and service economies of the minority world. Perhaps surprisingly, this wide-ranging field of practice has not been subject to much attention by commentators on the impact of behavioural science on public policy. Sometimes, as is the case with Mindapples, there is indeed a reluctance to be associated with the behaviour change agenda, as – for this organisation at least – this would involve unwarranted judgements about what people want to do, rather than giving them the tools to achieve their own goals (Andy Gibson, CEO of Mindapples, interviewed 5 March 2013).

Given the dominance of work in contemporary human life worldwide, and the pivotal place held by paid work in current conceptualisations of active citizenship within Western neoliberal democracies, workplace wellbeing programmes deserve academic scrutiny. Despite the differences outlined above, we argue they share three main characteristics with public sector policies aimed at changing citizens' behaviour. First, they are forms of emotional governance aimed at managing emotions, cultivating new forms of self-government, and creating what are deemed to be positive affects, individually, in terms of workers' relationships with each other, and in workplace atmospheres or cultures. Second, they draw on relatively novel and popularised behavioural, psychological and neuroscientific insights – coupling these with management and organisational theories which have long been influential in workplace training, performance management and institutional change. Finally, like the behaviour change agenda, wellbeing programmes have provoked critique from geographers who have sought to highlight their alignment with a neoliberal mode of economic organisation in which the state provision of collective welfare is replaced with a concern for individualised wellbeing (Pykett, 2015).

Dwelling for a moment on critical appraisals of a public policy emphasis on wellbeing, we can identify a number of common themes. Many tackle wellbeing schemes with reference to wider developments in 'happiness economics', including the initiation of Gross National Happiness as a national policy goal to be compared internationally and constantly improved. Sociologists have therefore drawn our attention to the way in which happiness and the optimisation of positivity have become new objects of governmentality which rely on the cultivation of entrepreneurial forms of subjectivity (Greco and Stenner, 2013; Binkley, 2014). As Will Davies (2011: 71) has noted, the very *pursuit* of happiness is essential to the functioning of a neoliberal economy in which perpetual growth, as well as the 'growth mindset' commonly evoked by workplace wellbeing training providers, is the goal. Actual achievement of happiness would be anathema within this economic framework, since a totally satisfied population would likely destroy consumer demand. The reduction of wellbeing to a form of economic *capital* or personal resource to be exploited, self-governed and maximised – for critics such as Sam Binkley, points towards the new ways in which the irrational, intimate and emotional self is being governed at a distance in a manner which determines the very *possibility* of subjectivity within a neoliberal framework. Other commentators too have found the policy emphasis on happiness and subjective wellbeing to be potentially suffocating and alienating, highlighting the blanket 'therapeutisation' of culture and of state functions (Nolan, 1998; Ecclestone, 2012; Furedi, 2014) and as a threat to democracy, mollifying potentially quarrelsome citizens. More specifically, the ideas of positive psychology which underpin much workplace wellbeing training have been dismissed as a mask for economic restructuring, rendering workers responsible for the economic fate of their companies or organisations, and demanding forms of gendered emotional labour which lead to the alienation of people from themselves in the blind pursuit of growth, performance, competiveness and profit (e.g. Hochschild, 2003; Ehrenreich, 2009).

These are serious criticisms, and ones which inform the analysis offered in the remainder of the chapter. In addition, we argue that, from a geographical perspective, it is also important to pay attention to the way in which both the problems to be achieved through behaviour change and the diverse goals of workplaces – be they social, economic, environmental, cultural, political – get reframed at the scale of not just the individual (in this case, the worker) but also the body. In workplace wellbeing programmes, the brain becomes the key object of governance, the goal being to cultivate forms of emotional subjectivity amongst workers which will align their everyday habits, relationships and performances with the goals of the organisation. So, too, workplaces and their organisational cultures become the key spaces in which economic successes and failures are played out, with the effect of taking them out of the specific historical and geographical context in which particular industries and sectors find themselves. With these reservations in mind, the next section focuses on the development and evaluation of the authors' own workplace intervention programme. This was aimed at improving participants' knowledge of behavioural science through a form of mindfulness practice which is becoming popular in workplace wellbeing initiatives in the UK and elsewhere. Workplace wellbeing programmes and behaviour change public policy interventions alike seek to intervene in people's impulsive, habitual and automatic decisions and behaviours. By contrast, the research discussed here sought to explore the potential of less individualistic and more group-based workplace mindfulness practice to provide one possible means to encourage people to *pay attention* to these very emotional drivers and behavioural cues of decision making. At the same time, it opened up a *discursive environment* in which workers might collectively reflect and learn about human behaviour and its contextual drivers in more empowering ways than offered by behavioural policies which target only the subconscious drivers of decisions (see also Rowson, 2011). As such, by combining mindfulness practice and learning about behavioural insights, our intervention has tried to set out an alternative and potentially more progressive means to shape policy makers' engagements with behavioural science knowledge, from a critical stance.

Combining behaviour change and mindfulness in public policy making

Policy makers themselves are of course subject to the same cognitive biases and impulses as citizens, but little attention has been paid to civil servants' own engagements with behavioural science knowledge. This section considers how the engagements of civil servants with behaviour change knowledge are emotional and embodied in nature, in light of an increasing recognition of the centrality of feeling and emotion to public sector work highlighted especially in chapters by Anderson (Chapter 6), Dobson (Chapter 8) and Jupp (Chapter 10) in this volume. These people are at once 'post-rational' humans, citizens, workers, public servants and responsible for the implementation of behaviour change initiatives themselves. One way in which civil servants are now being recognised as emotional subjects themselves is through recent initiatives introducing mindfulness practice within government, with MPs having been offered mindfulness training since 2013 by the Oxford Mindfulness Centre in order to support parliamentarians in making better decisions and responding more appropriately to situations. Mindfulness is a form of secular therapeutic practice, having been adapted from its original Buddhist roots by medic Jon Kabat-Zinn in the USA during the 1980s. It refers to the development of 'present-centred non-judgmental awareness" (Williams and Kabat-Zinn, 2013: 6) and involves people training their attention through breathing mediation, 'body scans' which focus the mind's attention on specific body parts, and the direction of thoughts and feelings in order to bring a focus on the present moment. Mindfulness is also aimed at supporting people to better understand the mental, emotional and environmental drivers of experience without being reactive to them, whilst rejecting normative judgements about their observations. In the UK, mindfulness programmes are now available as therapeutic treatment for particular mental illnesses within the NHS, and it has been offered in schools, workplaces and within government as a form of stress reduction. In 2014 an All-Party Parliamentary Group on Mindfulness was established to develop policy recommendations for government based on existing evidence and practice of mindfulness. It is also, as noted above, increasingly widespread in the delivery of workplace wellbeing programmes, offered to employees of several global corporations. Our programme differed from these more individualistic and therapeutic forms of mindfulness by emphasising critical awareness of the role of cognitive biases in decision making, by taking a group-based deliberative approach and by explicitly tackling the wider contextual drivers of human and social practices.

As part of our participatory research intervention, the development of a Mindfulness, Behaviour Change and Engagement in Public Policy programme, we conducted a pre- and post-intervention survey. This demonstrated that participation in the MBCEPP programme led to a statistically significant increase (at the 95% confidence level) in the level of participants' knowledge of each of ten behavioural insights set out in the survey; see Lilley et al. (2014) for the full evaluation report. In terms of the impacts of the programme on mindfulness, there was only a statistically significant increase in one of five facets of mindfulness, that of 'awareness'. In the qualitative interviews, participants said that they had felt a moderate increase in all five traits (observing, describing, non-reacting, awareness, non-judging). We found that the programme had several potentially beneficial effects for the civil servants. First, it enabled participants to reflect on the nature and purpose of policy, to consider ways in which behavioural insights might improve the *effectiveness* of policies:

> [w]e are so obsessed with process, and the way we do policy and legislation it seems that our mantra is that people will just do it because we say. They have just got to do it! And even if you use a law, it's not always effective because people resent it or people only do the bare minimum . . . It just seems that we need to be more sensitive and it's more complex than we think.
>
> (MBCEPP Participant 1)

There is a sense here of wanting to work in emotionally sensitive ways *with* citizens, engaging with people to achieve more long-term policy outcomes. There is recognition that telling people what to do through legislation may not be the most effective way to design policies.

Second, the programme led participants to question their own *ethical* perspectives, namely the non-alignment of their personal values with those of the government. One participant observed that working for government could sometimes feel rather corporate, driven by targets, results and standardisation. They expressed a feeling that working practices needed to be more driven by cooperation, mutual help and recognition of staff as assets to be valued and trusted. They felt that the MBCEPP programme improved their empathy with others in general. Moreover, it enabled them to re-evaluate their own commitments and to consider ways in which their job as a civil servant should be much more focused on serving the public and developing meaningful relationships with citizens in order to serve their needs:

> You know, I'm here to, I wanted to do a public service. This is what I want to do. And I don't think we're sensitive enough to the public and the public's needs . . . We spend so much time doing the strategy, writing it down, producing a lovely document . . . What we should be doing is investing in relationships, and we should be talking to people, and we should be out there really getting to grips, and understanding what people are going through.
>
> (MBCEPP Participant 1)

Third, bringing together mindfulness practice and information on behavioural insights was found by participants to be *empowering*, because it enabled them to become more aware of the emotional, habitual and automatic cues for their own behaviour. Although not intended as a therapeutic intervention, participants reported personal benefits in terms of stress reduction, a sense of calmness, better enjoyment of everyday activities or tasks, the ability to listen to others more effectively and to cope and deal with difficult work-based situations better, and to better respond to change. The group nature of the MBCEPP sessions was found to be rewarding for the participants, who stated that sharing their experiences with others in their organisation enabled them to understand themselves better and reduced any sense of hierarchy in the organisation. The participants were also encouraged that their employer was creating a space and time for them to focus on their own wellbeing, providing them with the sense that their emotional health was deemed important. One respondent reported that this aspect was particularly important in the context of public sector cuts and the pressures this brought to staff:

> There's a lot of people who were under a lot of stress because of what they have to deliver and cuts in resources, staff cuts. You know, more and more we are expected to do more and more with less resource. So I think that the course can help people to just entitle them to a general sort of health and wellbeing. I think that it can help people to manage their stress levels and how they manage things generally.
>
> (MBCEPP Participant 3)

We observed that, through the development of mindfulness traits including awareness of self and others, participants seemed to express support for behaviour change policies that were empowering, and felt that learning mindfulness practice alongside knowledge about behavioural insights was a useful way to enable people to assume more control over their behaviours and retain a sense of autonomy. But the programme did not just have effects in terms of self-knowledge. Participants also reported that knowledge of the minds and behaviours of others was significant. One participant stated that, in her work on tackling educational inequality, understanding policy makers' and teachers' own biases would be central to establishing educational policies and classroom practice which could mindfully tackle unconscious forms of discrimination. She saw this as distinctly preferable to what she perceived as the apparent ethos of subconscious 'trickery' of the nudge approach precisely because it did not circumvent people's ability to pay attention to and learn from their own biases. Several participants took from the MBCEPP programme a core focus on 'de-biasing', suggesting that both for themselves in their interactions with colleagues, stakeholders and services users, and in their aspirations for citizens too, they saw a central place for *learning* about the self as a form of empowerment. Senses of collectivism and of learning about others and the wider environmental context in which behaviours are shaped, often in constrained or unequal circumstances, were unexpected findings of our research. To a degree, these findings address the common criticism that mindfulness is a self-regarding practice ignorant of social relationships and contextualised, worldly knowledge and instead suggest there is at least potential for it to provide a means for more critical, ethical and other-regarding practice – though clearly whether this is possible will depend on the purpose of specific courses, the intentions of whoever commissions a course, and the perspective and knowledge of the mindfulness teacher.

Conclusion: governing the emotions of the governors

There are of course limitations and potential drawbacks to the kind of experimental intervention described here. The first is methodological; our evaluation of the MBCEPP programme relies on self-reported perceptions and experiences of the participants, as discerned by researchers who were also involved in delivering the programme. It is therefore difficult to draw any causal links between the programme and its effects. However, our aim in this research has not been narrowly to measure the cause and effect of the intervention but to investigate its potential to address the significant political and ethical concerns raised by efforts to change behavioural and emotional forms of governance as outlined above – namely that these should be transparent, empowering, open to deliberation, contextually sensitive, mindful of the potential for affecting already marginalised citizens, and encourage emotional reflexivity. The second obvious methodological limitation is that, to see the effects of this programme, the research would need to be longitudinal. Our feedback/feed-forward workshop went some way to addressing this, as will future follow-up interviews after six months. In order to fully understand the potential of combining mindfulness practice with education on behaviour change it would be useful to follow up how participation in the programme may have influenced these civil servants, their work on specific projects and their interactions with policy stakeholders and citizens in the much longer term.

In terms of potential drawbacks, there is a risk that pursuing mindfulness practice in a public sector workplace setting is seen as a cynical attempt to govern the intimate and personal emotions of civil servants; to train them in therapeutic techniques of self-management; to improve their effectiveness and productivity whilst at the same providing fewer resources to deliver core policy and services; to stave off conflict; to soothe and appease potential resistance; and to align the personal values, beliefs and behaviours of civil servants with the organisational goals of the government. For some critics, the MBCEPP programme will be considered as a strategic deployment of emotional intelligence and wellbeing in the service of dubious neoliberal interests which threaten the very existence of the state. The voluntary nature of participation in the research intervention meant that we were unlikely to discover such critiques amongst participants, since they were already positively predisposed to taking part. Yet these are precisely the kinds of issues and problems that are being discussed by mindfulness teachers, responding to what has been described as the dark side of 'McMindfulness' and as a passing corporate fad (Purser and Loy, 2013). Of course, the experience and perspective of specific mindfulness teachers will have a distinct bearing on whether the form of practice will be appeasing or more challenging and empowering.

The position we have taken in this chapter has been to forge a research path which accepts these critiques at the same time as trying to develop a potentially constructive and alternative approach to the implementation and application of behavioural insights by and with civil servants. This is a path which many will find too compromised or pragmatic. But we argue for the importance of maintaining a critical stance towards the interrelated ventures of behaviour change and workplace wellbeing, and through our ongoing development of the MBCEPP programme are seeking to establish a viable means to challenge the reduction of complex social, political and economic problems to individual behaviours in ways which carry currency in the current political climate.

We have set out to do this in ways which prioritise an approach to behaviour change as effective in the longer term, ethical and empowering. This requires some ambitious and experimental methodologies which respond to the experiences of research participants and which retain a certain scepticism towards research methods which limit themselves to narrow definitions of evidence, such as randomised controlled trials and psychometric tests. Rather, we argue that our methodology has opened up a space for unexpected findings to appear, namely that teaching civil servants about behaviour change through mindfulness brings our attention to the stresses facing civil servants in their working lives. The mindfulness teacher explicitly did not focus on stress during the course, and it was not intended as a therapeutic intervention, yet it seemed to benefit participants in this way. Furthermore, our approach resolutely did not promote behaviour change or workplace wellbeing as straightforward public goods beyond question. In contrast, we have tried to sketch out a specific context in which behaviour change and workplace wellbeing become viable as a solution to a quite specific sense of ineffectiveness in public policy work, constrained by a lack of financial resource and a culture of (de-humanising) standardised performance management. Whilst mindfulness and behavioural insights can be used uncritically in the 'therapeutisation' of culture and the governmentalisation of the emotions, this research suggests that, when understood *in context* rather than in a social, political and economic vacuum, these kinds of interventions in our emotional literacy can bring to conscious awareness a collective set of concerns about social and political change which may be potentially fruitful.

Notes

1 ESRC award ES/L003082/1. We would like to thank the research participants at the Welsh Government for their participation.
2 UK Network for Mindfulness-Based Teachers Good practice guidelines for teaching mindfulness-based courses, at: http://mindfulnessteachersuk.org.uk/#guidelines (accessed 21 April 2015).
3 MINDSPACE stands for Messenger, Incentives, Norms, Defaults, Salience, Priming, Affect, Commitments, Ego (Dolan et al., 2010).
4 As well as MINDSPACE, mnemonics include NUDGES (iNcentives, Understanding mappings, Defaults, Give feedback, Expect error, Structure choices) (Thaler and Sunstein, 2008) and EAST (Easy, Attractive, Social and Timely) (Service et al., 2014).
5 www.mindapples.org/services/training (accessed 19 March 2015).

References

Baer, R.A., Smith, G.T., Hopkins, J., Krietemeyer, J. and Toney, L. (2006) 'Using self-report assessment methods to explore facets of mindfulness', *Assessment*, 13(1): 27–45.

Binkley, S. (2014) *Happiness as enterprise: an essay on neoliberal life*. SUNY Press, Albany, NY.

Bissell, D. (2011) 'Thinking habits for uncertain subjects: movement, stillness and susceptibility', *Environment and Planning A*, 43(11): 2649–65.

BIT (Behavioural Insights Team) (2015) 'The global spread of behavioural insights: conditions for success of a Central Unit', 30 January 2015, at: www.behaviouralinsights. co.uk/blogpost/global-spread-behavioural-insights-conditions-success-central-unit (accessed 21 June 2016).

Chatterton, T. (2011) *An introduction to thinking about 'energy behaviour': a multi-model approach*. Department of Energy and Climate Change, London.

Cromby, J. and Willis, M.E.H. (2014) 'Nudging into subjectification: governmentality and psychometrics', *Critical Social Policy*, 34(2): 241–59.

Darnton, A. (2008) *GSR behaviour change knowledge review. Reference report: an overview of behaviour change models and their uses*, at: www.civilservice.gov.uk/wp-content/uploads/2011/09/Behaviour_change_reference_report_tcm6-9697.pdf (accessed 21 April 2015).

Davies, W. (2011) 'The political economy of unhappiness', *New Left Review*, 71 (Sept.–Oct.): 65–80.

Dolan, P., Hallsworth, M., Halpern, D., King, D. and Vlaev, I. (2010) *Mindspace: influencing behaviour through public policy*. Institute for Government and Cabinet Office, London.

Ecclestone, K. (2012) 'From emotional and psychological well-being to character education: challenging policy discourses of behavioural science and "vulnerability"', *Research Papers in Education*, 27(4): 463–80.

Ehrenreich, B. (2009) *Smile or die: how positive thinking fooled America and the world*. Granta Publications, London.

Friedli, L. and Stearn, R. (2015) 'Positive affect as coercive strategy: conditionality, activation and the role of psychology in UK government workfare programmes', *Medical Humanities*, 41: 40–7.

Furedi, F. (2014) 'Mindfulness is a fad, not a revolution', *Times Education Supplement*, 18 April 2014.

Greco, M. and Stenner, P. (2013) 'Happiness and the art of life: diagnosing the psychopolitics of wellbeing', *Health, Culture and Society*, 5(1): 1–19.

Hafenbrack, A.C., Kinias, Z. and Barsade, S.G. (2013) 'Debiasing the mind through mediation: mindfulness and the sunk-cost bias', *Psychological Science*, 25(2): 369–76.

Hallsworth, M. and Sanders, M. (forthcoming) 'Recent developments in applying behavioural science to public policy', in Spotswood, F (ed.) *Governance and Behaviour Change*. Policy Press, Bristol.

Hochschild, A. (2003 [1983]) *The managed heart: commercialization of human feeling*. University of California Press, London.

John, P., Cotterill, S., Richardson, L., Moseley, A., Stoker, G., Wales, C. and Smith, G. (2011) *Nudge, nudge, think, think*. Bloomsbury, London.

Jones, R., Pykett, J. and Whitehead, M. (2011) 'Governing temptation: changing behaviour in an age of libertarian paternalism', *Progress in Human Geography*, 35(4): 483–501.

Jones, R., Pykett, J. and Whitehead, M. (2013) *Changing behaviours: on the rise of the psychological state*. Edward Elgar, Cheltenham.

Kabat-Zinn, J. (2005) *Coming to our senses: healing ourselves and the world through mindfulness*. Piatkus Books, London.

Kahneman, D. (2011) *Thinking fast and slow.* Penguin, London.

Lea, J., Cadman, L. and Philo, C. (2014) 'Changing the habits of a lifetime? Mindfulness meditation and habitual geographies', *Cultural Geographies*, 22(1): 49–65.

Leggett, W. (2014) 'The politics of behaviour change: nudge, neoliberalism and the state', *Policy and Politics*, 2(1): 3–19.

Le Grand, J. and New, B. (2015) *Government paternalism: nanny state or helpful friend?* Princeton University Press, Princeton, NJ.

Lilley, R., Whitehead, M., Howell, R., Jones, R. and Pykett, J. (2014) *Mindfulness behaviour change and engagement in public policy: an evaluation*, at: https://changing behaviours. wordpress.com/2014/10/10/mindfulness-and-behaviour-change-an-evaluation/ (accessed 21 June 2016).

National Institute for Health and Clinical Excellence (NICE) (2007) *Behaviour change: the principles for effective interventions*. National Institute for Health and Clinical Excellence, Manchester.

Nolan, J.L. (1998) *The therapeutic state: justifying government at century's end.* New York University Press, London.

Purser, R. and Loy, D. (2013) 'Beyond McMindfulness', *Huffington Post*, 1 July 2013, at: www.huffingtonpost.com/ron-purser/beyond-mcmindfulness_b_3519289.html (accessed 21 June 2016).

Pykett, J. (2012) 'The new maternal state: the gendered politics of governing through behaviour change', *Antipode*, 44(1): 217–23.

Pykett, J. (2015) *Brain culture: shaping policy through neuroscience.* Policy Press, Bristol.

Rose, N. (1989) *Governing the soul: the shaping of the private self*. Routledge, London.

Rowson, J. (2011) *Transforming behaviour change: beyond nudge and neuromania*. RSA, London.

Schüll, N. and Zaloom, C. (2011) 'The shortsighted brain: neuroeconomics and the governance of choice in time', *Social Studies of Science*, 41(4): 515–38.

Service, O., Hallsworth, M., Halpern, D., Algate, F., Gallagher, R., Nguyen, S., Ruda, S., Sanders, M. with Pelenur, M., Gyani, A., Harper, H., Reinhard, J. and Kirkman, E. (2014) *EAST: four simple ways to apply behavioural insights*. Behavioural Insights Team, London.

Thaler, R.H. and Sunstein, C.R. (2008) *Nudge: improving decisions about health, wealth and happiness.* Yale University Press, London.

Wells, P. (2010) 'A nudge one way, a nudge the other: libertarian paternalism as political strategy', *People, Place and Policy*, 4(3): 111–18.

Whitehead, M., Jones, R. and Pykett, J. (2011) 'Governing irrationality, or a more than rational government? Reflections on the rescientisation of decision making in British public policy', *Environment and Planning A*, 43(12): 2819–37.

Whitehead, M., Jones, R., Pykett, J. and Welsh, M. (2012) 'Geography, libertarian paternalism and neuro-politics in the UK', *Geographical Journal*, 178(4): 302–7.

Whitehead, M., Jones, R., Howell, R., Lilley, R. and Pykett, J. (2014) *Nudging all over the world: assessing the global impact of the behavioural sciences on public policy*, at: https://changingbehaviours.wordpress.com/2014/09/05/nudging-all-over-the-world-the-final-report/ (accessed 21 June 2016).

Williams, J.M.G. and Kabat-Zinn, J. (2013) *Mindfulness: diverse perspectives on its meaning, origins and applications*. Routledge, London.

6   The sentimental civil servant

Rosie Anderson

Introduction

In their seminal ethnographic study of the UK senior civil service, Mark Bevir and Rod Rhodes argued that 'it is only through the micro-level analysis of minister and public servants at work that we can understand how governance changes' (Bevir and Rhodes, 2006: 109). By employing this micro-level analysis, how policy actually gets made becomes foregrounded – the practice of policy – in all its messiness and confusion. This inevitably leads the researcher to re-examine the way policy workers think about and try to make sense of their work. For Bevir and Rhodes, this was an important tool for reassessing conventional narratives about how governance happens, particularly its supposed linearity and centralised coherence. This chapter argues that paying close attention to the micro-interactions of civil servants involved in policy discussions also reveals a complex set of cultural expectations around the behaviour, language and bodies of civil servants.

What has been referred to as the 'practice turn' in policy analysis (Freeman and Maybin, 2011; Wagenaar, 2004) has highlighted the importance of the micro-interactions of governance in understanding how policy gets made and has sense made of it. Perhaps inevitably, this concern for understanding the fine grain of policy work has led to an incipient but growing interest in the affective and relational dimension of governance, particularly with reference to partici-patory approaches to decision making (Welch, 1997; Fischer, 2009; Hunter, 2010). Despite this, the interpersonal, physical presentation and presence of civil servants (or indeed anyone) has received strangely little attention among scholars of policy and politics. Generally researchers have used the traditional artefacts of policy work to explore the emotional side of governance and policy work, notably opinion polls and electoral counts (Marcus, 2000, 2003; Westen, 2008) as well as the texts of policy decisions or the content of evidence (Hardill and Mills, 2013). This emphasis on text and discourse in the study of emotion in policy making perhaps reflects the models of policy making commonly employed by scholars, namely that policy and politics more generally is a deliberative, argumentative process which is 'made of language' (Majone, 1989: 1). The purpose of such deliberation is to reason, through argument or analysis, towards a mutually agreed decision. Any behaviours or practices deemed to be disruptive to the process of posing and responding to argument and deliberation have tended to be seen as 'non-productive sidetracks' (Fischer, 2003: 177). 'The emotional' has been analysed by Goodwin et al. (2001), Fischer (2003), Barnes (2008) and Newman (2012) as a significant but neglected facet of that reasoning process, with the emphasis very much upon integrating emotion – the 'non-rational' – into a deliberative framework, Newman in particular concluding that this may not always be possible. Other scholars have looked for philosophical and ontological frameworks that emphasise the primacy of feelings in their broadest sense to arriving at judgements, and therefore by extension reasoned decisions such as policy making. These draw upon the idea of a somatic, empathic response to phenomena in the world, and particularly others' experiences of the world, as valuable knowledge of a particularly 'emotional' sort (Nussbaum, 2003, 2013; Krause, 2008), a form of intelligence about the world.

It is perhaps no coincidence that surprisingly little interrogation of the everyday meaning of emotion to policy participants has taken place among policy and politics scholars. A return to Bevir and Rhodes' emphasis on observing practice – including the relational, the physical and the embodied – to avoid importing assumptions from one's received understanding of life in government departments (or any place that policy takes place) is essential if we are to understand the meaning of emotion in policy. The research presented here set out to try to take emotion as the starting point for its inquiry into the practices of political work, rather than as something which had to be accounted for within any particular model of what policy making is. In turn it demanded that some assumptions about the linear and purposive nature of policy work be interrogated.

This chapter is an account of one attempt to build a picture of the everyday meanings and practices of emotion in the context of policy making. It focuses in particular on civil servants as people who were described by informants as having a particular affinity with the non-emotional, and who at the same time displayed complex and marked shifts in physical and vocal emotional self-presentation depending on their physical and figurative relationship to the people and issues they encountered.

After an overview of the research project and a consideration of the everyday meaning of emotion to informants, the chapter describes the particular relationship to emotion that competent civil servants were expected to display. It also interrogates the implied intertwining of these emotional participant roles with different narratives about morality and governance. In particular, it proposes that the philosophical concept of 'sentiment' – the use of one's recognition of another's experience of a phenomenon to understand one's own moral response to it – was closely tied to the physicality of being morally present in the room with another and those present having been mutually acknowledged as full persons.

Finally, an analysis of these different 'emotionalities' and the shifts that took place is proposed, looking at how civil servants moved through socially constructed spaces for policy making, changing their stance in relation to emotional knowledge and other participants primarily through their use of space, gesture and physicality.

Research in interesting times: background and methods

This chapter is based upon a doctoral research project looking at the meaning and practice of emotion in policy making. Fieldwork took place between January 2012 and April 2013. I entered the field by working part-time on a voluntary basis for an NGO I will refer to as 'the Partnership' which was involved in a range of policy areas across the UK government, the devolved administration at Holyrood in Scotland, local government and occasionally EU departments, with the broad aim of combating poverty. I provided unpaid policy officer-type support, a role I had several years' professional experience in prior to starting my doctorate. My duties consisted of supporting the organisation's activities across its programme of policy work, the majority of which was focused on the devolved administration in Scotland. The methods used combined non-participant observation as I shadowed my new colleagues and 'learned the ropes', participant observation as I went about my work as a policy officer and formal, recorded interviews with key participants. This yielded a range of different types of data, specifically: extensive field notes from participant and non-participant observation; transcripts and recordings of interviews; various written documents such as briefing papers, minutes of meetings and reports, including those I wrote myself; and the electronic communications exchanged via email or social networking sites in the course of working life.

My research was by its nature potentially sensitive and reputationally risky for my informants, and the somewhat ad-hoc nature of the work of policy making made issues around consent and confidentiality highly salient but also complex, practically speaking. Scotland is a relatively small legislature and many of my informants inevitably knew about each other's participation in my research. I have used pseudonyms for all organisations and individuals throughout this chapter, and only given enough description of people's role and organisation to contextualise them and their work within a credible place and time.

Originally it was intended that among other duties I would provide intensive support to a particular forum which sought to bring together key stakeholders in an ongoing discussion explicitly designed to foster close working relationships. This forum met twice a year but also consisted of sub-groups that worked on specific aspects of policy throughout the year. Across its work, the NGO conceived of its key stakeholders as comprising three groups: people directly experiencing poverty; NGOs and similar national advocacy groups; and civil servants and parliamentarians. This way of categorising the forum's participants was extremely consistent across all participants and across time. A programme of targeted repeat interviews was proposed at the beginning of fieldwork which would involve two forum participants from each category.

However, the wider circumstances of policy work in the UK forced several changes of plan. About six months in it was known that this forum was not going to have its funding renewed. A couple of months later it became clear that not just this forum but the whole programme of work it formed a part of was going to have to be abandoned as no new funding could be found. At this point, the colleague in charge of the programme announced that she would leave before the end of the funding cycle and the organisational support for the project effectively collapsed. Outside the organisation, a ministerial reshuffle took place and ongoing cuts dominated Scottish Government and parliamentary business. For obvious reasons, this changed the activities of the forums and greatly curtailed the paid participants' ability to fully engage with the overall programme. It perhaps had the most marked effect upon the contact time I had with civil servants who were involved in the work of the forums, who were generally drawn from the permanent middle ranks of a range of Scottish Government departments and working on strategic policy reviews, proposals for legislation as part of a 'Bill team' or similar tasks. My experience of this category of participants is of peripheral but hugely significant actors in the policy-making spaces we all inhabited at the time, an experience which was typical under the circumstances. This chapter is therefore written very much from the perspective of someone watching, or maybe assisting in, the performance of emotionality by the civil servants involved in the forum. Nevertheless, the experience of the civil servants that I did engage with suggests that there is a rich and complex culture around emotion in the civil service which deserves greater attention.

Emotion as knowledge and being-in-the-world

I had expected to find that my informants would identify certain behaviours as stereotypical, and that certain people would be sanctioned to behave more 'emotionally' than others. Perhaps, I surmised, women might be 'meant' to be more emotional than men, older people would have been expected to have been less emotionally demonstrative or sensitive, and so on. What surprised me about the way emotion was articulated to me was the absolute primacy of function and occupation in determining who was considered 'emotional'. By my interlocutors' own insistence they were grouped into set categories of stakeholders, each with different 'roles' that symbolised something about their relationship with emotion. One of the most direct descriptions of these roles was provided by an activist, Carla, about six months into my fieldwork. We had met through the Partnership's outreach work. She was a long-standing community activist to whom the Partnership staff had devoted considerable efforts in order to enable her to participate. I asked Carla for her practical input in spotting the 'emotional' and she was very clear and unhesitating in her reply: I needed to watch the activists – the 'people experiencing poverty'. So, I asked, the other people involved don't 'do' emotion, then? She affirmed this:

> You'll get more truth from the community activists . . . I think they know from experience, where other people only know from what they've read or heard, but I think the actual person who experiences it can tell it proper.

Carla made it clear that people who worked for NGOs or in Parliament or Government would not be bringing any emotional content to the events I attended;

'I think they just go by what they have written down,' she said. She believed that this was the only way that people who worked as 'professional' policy makers had access to the 'reality' of the issues they talked about; this was in fact a hallmark of their professional status and part of what *made them what they were*. Emotion was not just a set of behaviours; it was a way of being-in-the-world which certain embodied experiences could provide, and others by definition could not.

Emotional knowledge and morality

The first time I personally became aware of the divergence in participants' behaviours and expectations around emotion in the Partnership's work was at the first annual showcase event that I attended. This came at the end of the first three months of my fieldwork. I attended as a participant observer but in a very low-key way; I was an independent delegate at a conference which numbered around two hundred. My understanding was the event was intended to demonstrate the work that the Partnership was doing around advocacy and 'voice' and the effect on policy agendas that impacted on poverty in Scotland. Delegates came from the three major stakeholder groups the Partnership identified in its strategy documents: Parliament and Government; NGOs and similar civil society groups; and 'grassroots' volunteers. Those in this last group were typically the speakers in the 'evidence sessions' convened by the Partnership, which aimed to bring policy- and decision makers into contact with the experiences of people living in poverty as a means to influence policy. Through observing and discussing these differences in presentation I hoped to begin to analyse the way emotion, knowledge and participant-role enmeshed in the context of the forums.

As it turned out, my experiences during these two days provided only provisional answers and in fact generated several complex questions, particularly around the gap between professional role and the person in a more general sense. One evidence session in particular left me uneasy and unable to explain what exactly had taken place at a practical, policy-influencing level. It followed the same structure as other evidence sessions – a community or 'grassroots' expert witness and a Government or civil service representative were introduced by a chair, each gave a brief outline of their take on the matter, and the discussion was opened up to the floor.

After two rather monotone presentations, the chair opened up the discussion to the small audience. From there things became increasingly fractious. Eventually a woman who identified as an activist stood up and delivered a long, very aggressive and rather fragmented monologue, personally attacking the civil servant and all his colleagues for being lazy, callous and overpaid and indirectly responsible for several deaths she was aware of on her estate.

The effect on the rest of the room was fascinating. I felt scared by the raw anger on display from this woman. Throughout her torrent of accusations most people in the audience had been looking awkward and staring at the floor but, I was interested to note, 'equally they clap at the end'.

In contrast I had noted to myself that the Scottish Government civil servant had pretty much read from the latest Government position paper, and he was somehow curiously absent. I thought this might be because of the way he used his body and voice during his time in front of the audience, although I didn't prompt him with this observation during our subsequent interview. I noted while he was addressing the evidence session that his eyes rarely rested on anyone in particular; his voice was evenly pitched and paced, monotonous even; his shoulders were raised towards his ears in a tense rictus; his face was somehow mask-like and expressionless, and I noted his eyes and eyebrows were practically frozen.

I decided that there was something about who the civil servant was in the room with us that had contributed to this state of affairs, and from my own assessment of the situation I felt he had no other legitimate choices in terms of who he was. I realised I needed to talk to him more about what he thought he was doing in the way he had presented himself in that encounter and why the display of raw, personal vitriol from the audience had apparently so wrong-footed him. I went straight up to him after the chair had rather shakily thanked everyone and wrapped proceedings up. The way the civil servant dealt with this request suggested to me that there was a complex set of codes about his behaviour in these encounters that he and I would need to discuss:

> I ask if I can interview him next week sometime. He looks initially confused, then wary, then hands out his email. 'It's anonymous of course,' I say, and smile what I suppose is a sympathetically knowing smile. The physical effect on the guy is almost perceptible; his shoulders tumble downwards, his eyes both focus on me and come alive; it's almost like someone being revived from stasis.

Douglas – as I will refer to him here – had been acting out something by behaving the way he did, and now that he was not 'on-stage' he could step away from his policy-making role and towards the complexities of Douglas as a human being.

Emotion, knowledge and truth: exploring the anxieties of governance

I met Douglas a week after the annual showcase in a coffee shop halfway between my department in the University and his department in the Scottish Government. I discovered that he was a career civil servant who had been involved intermittently with the Partnership and the forum as a member of various Bill teams, drafting Government white papers. I explained to him that throughout the evidence session he and I had been present at I was intrigued not just by what everyone said, but also by the way they said it, their presence in the room. I asked him to explain what he felt had happened, and he brought up the idea of roles:

> It's almost like there are rules drawn up around these things, there are like patterns of behaviour, which [we] have learned, whether it be business which

is you know about controlling yourself and trying to get people to listen to you or whether it be . . . Yes, how to reflect that passion but also how to use it effectively and the like.

Douglas went on to explain the double-bind of being a policy 'professional'; 'I can't answer those questions [about someone's personal life] because they're individual experiences. I need to respond as an individual but I'm not *there* as an individual' (original emphasis).

Douglas was presenting the 'personal' as being in direct opposition to everything that he must embody as a 'public' figure, at the loss of his individual presence. His bodily and vocal behaviour at the conference was in some way a performance of this absence, the reading from Government documents a stratagem for faithfully suppressing his knowledge that wasn't rational professional knowledge; he was quite literally 'reading from a book', as Carla had put it.

At the same time, Douglas believed that he still cared about the people and the issues at hand, he still had emotional knowledge, and that it needed to be accessed to make good decisions. Indeed it was the only thing that could make rationality usable or good. When I asked him why he came to events like the one we'd just been at if he expected to get shouted at, he explained that such events enabled contact with people *experiencing* poverty – not studying poverty, or analysing poverty, or representing those in poverty – and that this was essential to uncovering 'reality', echoing Carla's language. 'It does make it more real,' he said. But he went on to explain that these encounters with 'reality' were essential for making not just effective but also just and normatively right decisions, for acting as a moral agent; 'otherwise you don't really get a sense of what you're doing, or why you're doing it'.

I felt that Douglas' statement begged a prior question about the 'truth' of emotion, mirroring Carla's insistence that it was only in the emotional testimony of activists that 'reality' was able to be understood by policy makers. It seemed that there were two aspects to Douglas' idea of emotion as truth. First, there was an ontological question about the nature of reality which is only accessible through the visceral and somatic experiences relayed in 'emotional' testimony; second, there was also a normative, epistemological one about the way we understand what is significant and important about reality which only that visceral experience can give us. This is the clue which tells us not just 'why you're doing something' but also at a very basic level 'what the significance is of what you're doing', i.e. the meaning of your actions in normative terms.

Douglas clearly believed that his presence in the forum was meant to stand for a particular relationship with 'reality', one grounded in rational, objective knowledge and transmissible through texts, numbers and other artefacts of governance. He could in fact sit at his desk and govern in this sense perfectly competently, or perhaps just look at emails submitted to public consultations, and stated that many of his colleagues preferred to do this, given the tendency for face-to-face encounters with the public to be somewhat bruising. But his representation of this state of affairs as an 'ivory tower' – a cliché as he admitted – implied to me that there was a moral problem with this sort of governance in that it lacked crucial information about the nature of social reality. In short, emotional knowledge of the world was necessary in order to understand the morality of one's decisions as a public servant. In order to do his job not just competently but justly Douglas needed to care about the way his work changed the world and the people in it, and experiencing these stories first hand and viscerally put him in touch with the emotional knowledge he needed.

Relating to reality: the embodied encounter in civil service practice

I got the opportunity to explore in some depth the reasoning behind another civil servant's engagement with the sort of first-person testimonies the Partnership specialised in through Craig, a member of a Strategic Review Team that had worked with the Partnership over some months. I had met him not long after the parliamentary summer recess in 2012, when he and a couple of colleagues came to feed back to some volunteer researchers how the evidence they had submitted to the Strategic Review had been considered by his team. I had been intrigued by Craig's incredibly deadpan, blank personal presentation in the feedback sessions, and my curiosity about how he understood the role he played as a representative of the state was only increased by the way he responded to my request for a coffee and an interview. When I asked if I could talk to Craig one-on-one about emotion and policy making, he just looked at me, face blank. After a short pause he said, 'You do know I'm a civil servant?', without changing his expression.

I asked him directly what he felt he got from coming to the events run by organisations like the Partnership; after all, he could just have read their submission and left it at that. Craig echoed his civil service colleagues' statements about 'emotional' testimony and truth:

Craig: I mean we have loads of statistics, the Department of Work and Pensions have loads of statistics and there are so many reports published and research and analysis and so on. And it's all valuable, it all highlights where the issues are in general terms. But it doesn't really tell you how to fix it. It doesn't tell you . . . [Pause]

Rosie: What do you mean?

Craig: It doesn't tell you the things that are getting in the way for customers. You know? It'll say the outcomes are bad for this client group or this part of the country, in terms of how you solve that if you don't actually go deeper and get kind of qualitative stuff, then you're just kind of using your imagination . . . That's why I'm quite passionate about hearing from the customers.

Craig's words were interesting to me because it appeared what he was saying was that good governance was not an exercise in empathetic imagination; that is to say, it was not a question of trying to imagine what it would be like if you were the person experiencing poverty. It was more direct, more visceral than that. It was about being attentive to your own feeling reactions to policy encounters.

It seemed to me that the implication of the Partnership's rationale for their events and the way civil servants spoke about their involvement was that there was something being transmitted through performance and presence which could not be conveyed through text or facts and figures. Jenny, one of the Partnership's board members and herself a policy worker for another NGO, elaborated on the Partnership staff's conviction that conveying the experience of poverty made for 'better' policy. She emphasised that including this knowledge was a way of getting at what really mattered and understanding what was really important about that experience:

> You know, where you've got a duff policy and policy makers need to know about it and they need that kind of direction so they can really sit up and take notice and really do something about it, you know?

When a policy is bad, Jenny's statement suggests, when it doesn't work, people suffer. Through not just knowing of their suffering by report or anecdote and intellectually disapproving of it but by experiencing the world in new ways via a somatic encounter, policy makers *feel* the badness of the policy and get a sense of the parts of it that must most urgently be put right.

I was surprised by this conceptualisation of morality in public administration, and surprised by its adoption by civil servants and NGO workers alike, because it seemed to be so much at odds with what I and they would most comfortably understand good governance to be founded on: deliberation and impartiality. The implications could be summarised as follows: in order to know what really matters about an issue, in order to really understand, you must really feel what others feel – not just know it – and you must really *care*. While a version of this principle was relayed to me whenever I asked 'professional' participants about the purpose of hearing testimony from people experiencing poverty, my interlocutors were unable to readily volunteer an explanation for why this was 'better' or 'good' and how it related to other narratives about 'good' public administration. Having exhausted explanations from the field, I spent some time reflecting on the normative argument being made for this special emotional form of knowledge as they described it.

The sentimental civil servant?

The knowledge contained and evoked by the 'emotional' was being described by my informants as offering an alternative morality to that of the knowledge of rationality and 'professionalism'. I began to wonder if these knowledges and positions were in fact enmeshed with two moral codes that apparently ran in tandem with each other. My informants, *including the civil servants themselves*, seemed to be saying that 'professionalism' was not enough to make morally good policy, and this compelled me to try to articulate this alternative moral framework.

Sentimentalism as such was never used as a concept within the forum or among its participants, but its central thesis fits these descriptions of the moral use of emotion in policy and politics: 'When we judge something is wrong, one or another of these emotions will ordinarily occur, and the judgement will be an expression of the underlying emotional disposition' (Prinz, 2006: 34). It is a thesis that has a venerable ancestry in the wider cultural context we operated within, most notably the thinking of Hume and the other British moralists. Krause (2008) has used this philosophical framework for thinking about the role of what she terms 'passion' in political work, with the aim of reconciling the liberal democratic emphasis on impartiality and reason with these compelling and very personal emotional reactions to the world and others. It has been argued by Richard Rorty that, as a form of moral epistemology, sentimentalism and Hume's moral sentiments can still be found in the very current and contentious political thinking around inalienable human rights, through an appeal to recognising sameness within the category of 'human' (Rorty, 2011).

I can certainly see a connection between this conceptualisation of human rights on a global and abstract level and the concept of rights in a specific, practical context concerning poverty in Scotland. My research has rather different ontological and epistemological underpinnings to both Rorty's and Krause's work, however, and I am concerned not to go too far in equating the folk sentimentalism I believe I observed in my fieldwork with the formal sentimentalism of either scholar. Furthermore, I would reformulate Rorty's 'sameness' as 'similarity', since I am not sure that empathic sameness was quite a fair representation of what the civil servants told me they were experiencing with those who testified at the Partnership's events. However, arguably the moral epistemological process is the same in both cases: for others to suffer is an outrage to their humanity. Outside of the Partnership's forum, one may or may not be intellectually convinced of this idea of 'truth' and moral sense, but inside it this sentimentalism – or some version of it – appeared to be an extremely important factor in practising 'good' policy making.

On a practical level other people expected to *see* that you felt something. Civil servants themselves were particularly explicit about this perceived moral imperative to be emotionally present in a decision. Douglas' use of language – 'otherwise you don't really get a sense of what you're doing, or why you're doing it' – was interesting; without this emotional dimension to a decision it was inappropriate in some way, even potentially without meaning. The implication of this sentimentalism was that without being emotionally present – 'there but not there' – decisions were potentially immoral, monstrous even. Something that makes people whole was lacking, and it is a short step from being unemotional to potentially being in some way inhuman.

This posed a very practical challenge to civil servants in being physically present in the forums. The thing that was so directly contrasted with emotional knowledge – technical, received expertise – was considered important by civil servants, activists and NGO workers alike and a key component in the construction of an identity as a 'professional'. There was a particular emphasis placed by all participants on the rationality and therefore supposedly unemotional nature of civil servants, and that was a key part of their legitimacy and their role as participants. Since the emotional stance and the reasonable stance were viewed as being incommensurate, someone needed to quite literally be the voice of reason, or the social order of the forum would start to spin off its axis. Towards the end of my time with the Partnership Jenny, the Partnership board member mentioned above, reflected with me upon the discomfort civil servants displayed when confronted with situations that might require less structured interaction with activists and 'grassroots' perspectives. She speculated: 'Maybe it's that fear . . . Maybe fear's too strong a word, but it's like any of us going into the unknown, they're outwith their comfort zone.'

The implication was that, in order to be a convincing civil servant, individuals needed to learn to demonstrate an inhibition of the 'non-rational' relationships they may have with people and situations. Civil servants themselves were very explicit about this part of their work, and the discussions I had about their attendance at 'evidence' sessions such as the annual showcase event or the Strategic Review feedback session naturally led on to them explaining the way that their behaviour differentiated them from activists as 'professionals' and dictated the way they encountered emotional knowledge. Craig explained how he had learned to be a professional for himself:

> The civil servants, the idea was always to retain a certain level of impartiality and not to get . . . too drawn in . . . to the emotion of the situation. Em . . . Particularly because you might, em, end up ... making suggestions or making hints of commitments that you can't . . . or the ministers haven't . . . approved. You can't get cornered too much.

He and others from the Scottish Government also explained that while emotional testimony – first-person, lived experience – was invaluable in getting a feel for what was important, it was also suspect knowledge. Craig went on:

> That might be a specific instance or it might be a symptom of a wider issue, so . . . we have to make a judgement about that . . . There is always that thing about . . . being careful not to over-commit. Or . . . not to over-sympathise . . . [laughs] If that's not to sound too cruel!

Perhaps for this reason, my request to talk with civil servants about emotions in their working life was often almost incomprehensible to them at first – it was almost as if they were struggling to allocate my invitation a place in the scheme of their work. At the risk of ruining Craig's joke – 'you do know I'm a civil servant' – it works because it goes without saying – *everybody* here knows – that there are no emotions in the work he does. He didn't need to tell me that it's inappropriate for him to have an emotional dimension to what he does; he doesn't need to justify excluding emotional things from his working life. That side of his professional self simply *does not exist*, and it is *impossible*. So what would we have had to talk about? At the same time, it also works as a joke because of its absurdity, the absurdity of his situation.

Conclusion – avatars of the state

This chapter has described the way being and knowing in the world were split by participants in the Partnership's work into two incommensurable modes of knowledge: emotion and rationality. Over the course of this discussion I have documented how I explored the link between types of knowledge, legitimacy and role in the Partnership's policy forums, and in particular how 'professionalism' and those in 'professional' roles were closely identified with rational knowledge and the way this was enacted in the practices of the Partnership's institutions. This split and subsequent identification between roles and knowledge perform both a pragmatic or procedural and a symbolic function in the practices of policy making. At face value there is the construction of a code of behaviour which outwardly demonstrates the values of the professionalism expected of civil servants or other people who work for the public good rather than personal interests in policy. It therefore serves an ethical purpose and could be considered as a healthy boundary, psychically speaking. Preserving that distance, retaining that 'certain level of impartiality' as Craig put it, is perhaps necessary not only to maintain one's own effectiveness in doing one's job, but also in preserving a sense of one's own values and perspective at a more fundamental level by not becoming overwhelmed by another's experiences. And yet the visceral experience of another's experience was also regarded as fundamental to being a morally present policy worker, setting up a split within the civil servant and the policy-making process itself which was regarded as unbridgeable and sometimes difficult, dangerous and painful.

It was necessary and perhaps morally right for the participants in the Partnership's policy work to remind ourselves of the separateness of our experiences of the world; the abstracted, technical knowledge that civil servants brought to the Partnership's work was indeed tied to their wider identity as servants of a public good that existed in reference to ideas and institutions external to those interpersonal encounters. However, due to tensions between what the 'good' in public good refers to and the moral implications of the way one experiences the world that circulated among the Partnership's participants, we collectively found ourselves in a situation where 'the contradictory truth that things or persons can be black and white, good and bad *at the same time* seems unimaginable' (Fuchs, 2007, p. 382). Douglas' experiences at the annual showcase event were efficient at maintaining that split between the moral civil servant and the moral activist, but unsatisfactory from the pragmatic point of view of the Partnership's ability to impact policy through such events. The Partnership's explicit aim was to build dialogue and consensus on policy matters which impact people experiencing poverty, but my observations suggested that the way participants related to these large public events was in fact more about externally dramatising an internal anxiety about moral governance and political legitimacy.

Governance and politics more generally are widely exhorted to be less advers-arial, more collaborative and more 'empowering' by those in and outside the academy. Indeed, much of critical policy studies and the 'argumentative turn' within it is aimed at engineering and promoting the 'dialogue' that the Partnership itself espoused (e.g. Fischer and Forester, 1993; Wagenaar, 2011; Hajer, 2005; Healey, 1992). Participants' motivations for taking part in policy making are maybe more complex and conflicted than this deliberative democratic ideal can accommodate.

References

Barnes, M. (2008) 'Passionate participation: emotional experiences and expressions in deliberative forums', *Critical Social Policy*, 28(4): 461–81.

Bevir, M. and Rhodes, R.A.W. (2006) *Governance stories.* Routledge, London.

Fischer, F. (2003) *Reframing public policy: discursive politics and deliberative practices.* Oxford University Press, Oxford.

Fischer, F. (2009) *Democracy and expertise: reorienting policy inquiry.* Oxford University Press, Oxford.

Fischer, F. and Forester, J. (1993) *The argumentative turn in policy analysis and planning.* Duke University Press, Durham, NC.

Freeman, R. and Maybin, J. (2011) 'Documents, practices and policy', *Evidence and Policy: A Journal of Research, Debate and Practice*, 7(2): 155–70.

Fuchs, T. (2007) 'Fragmented selves: temporality and identity in borderline personality disorder', *Psychopathology*, 40(6): 379–87.

Goodwin, J., Jasper, J.M. and Polletta, F. (2001) *Passionate politics: emotions and social movements.* University of Chicago Press, Chicago, IL, and London.

Hajer, M.A. (2005) 'Setting the stage: a dramaturgy of policy deliberation', *Administration and Society*, 36(6): 624–47.

Hardill, I. and Mills, S. (2013) 'Enlivening evidence-based policy through embodiment and emotions', *Contemporary Social Science*, 8(3): 321–32.

Healey, P. (1992) 'Planning through debate: the communicative turn in planning theory', *Town Planning Review*, 63(2): 143.

Hunter, S. (2010) *Power, politics and the emotions.* Routledge, London.

Krause, S.R. (2008) *Civil passions: moral sentiment and democratic deliberation.* Princeton University Press, Princeton, NJ.

Majone, G. (1989) *Evidence, argument, and persuasion in the policy process.* Yale University Press, New Haven, CT.

Marcus, G.E. (2000) 'Emotions in politics', *Annual Review of Political Science*, 3(1): 221–50.

Marcus, G.E. (2003) 'The psychology of emotion and politics', in Sears, D.O., Huddy, L. and Jervis, R. (eds) *Oxford handbook of political psychology.* Oxford University Press, Oxford, pp. 182–221.

Newman, J. (2012) 'Beyond the deliberative subject? Problems of theory, method and critique in the turn to emotion and affect', *Critical Policy Studies*, 6(4): 465–79.

Nussbaum, M.C. (2003) *Upheavals of thought: the intelligence of emotions.* Cambridge University Press, Cambridge.

Nussbaum, M.C. (2013) *Political emotions: why love matters for justice.* Harvard University Press, Cambridge, MA.

Prinz, J. (2006) 'The emotional basis of moral judgments', *Philosophical Explorations*, 9(1): 29–43.

Rorty, R. (2011) 'Human rights, rationality and sentimentality', in Singh Rathore, A. and Cistelecan, A. (eds) *Wronging rights? Philosophical challenges for human rights*. Routledge, New Delhi and London, pp. 107–31.

Wagenaar, H. (2004) '"Knowing" the rules: administrative work as practice', *Public Administration Review*, 64(6): 643–56.

Wagenaar, H. (2011) *Meaning in action: interpretation and dialogue in policy analysis.* M.E. Sharpe, Armonk, NY.

Welch, D.D. (1997) 'Ruling with the heart: emotion-based public policy', *Southern California Interdisciplinary Law Journal*, 6: 55.

Westen, D. (2008) *Political brain: the role of emotion in deciding the fate of the nation.* PublicAffairs, New York.

Part III

Emotions in public services

7 Behaviourally, emotionally and socially 'problematic' students

Interrogating emotional governance as a form of exclusionary practice in schools

Jennifer Lea, Louise Holt and Sophie Bowlby

Introduction

The governance of emotions has become part of the educational experience of many, if not most, school pupils in England. Governance, as understood by Foucault (1991), is a particular form of encounter between states and subjects. Rather than acting directly on subjects, the state instead 'responsibilises' subjects, so that they turn on themselves to monitor, act and thus produce the *right kind* of selfhood. Schools, as the main site for the formation of future subjects, offer a realm in which this kind of responsibilisation can take place. While this might happen across a variety of registers, the one under discussion here is emotional governance. The *right kind* of emotional subjecthood, framed via a wider 'cultural and political "therapeutic" sensibility' (Ecclestone, 2010: 62), serves an immediate instrumental aim to create children to function as 'pupils' in the institutional context of the school (i.e. non-disruptive, conforming to disciplinary norms, able to learn) and is seen as increasingly important in its own right (as a fundamental part of individual wellbeing and happiness), as well as being positioned to be 'integral to personal development, life and work success' (Ecclestone, 2010: 63).

Such forms of emotional governance (in which pupils are encouraged to become responsible for the management and development of their emotional selves) have taken two main forms: first, curriculum interventions which encompass *all* pupils and, second, targeted interventions for children who are deemed to have abnormal emotions or experience difficulty in emotional expression and management (such as nurture groups). This chapter looks at the first of these, taking the case study of SEAL – or the Social and Emotional Aspects of Learning. Paying close attention to circle time (a significant mode of SEAL delivery), the chapter asks how far the SEAL curriculum, and its received mechanisms of delivery, might create forms of exclusion whereby *some* children are excluded from developing the *right kind* of emotional subjecthood; the chapter focuses on children who are designated as having Behavioural, Emotional and Social Difficulties (BESD).

The UK government definition of BESD suggests that pupils thus designated are 'withdrawn and isolated, disruptive or disturbing, hyperactive and lacking concentration, [with] immature social skills, presenting challenging behaviour arising from other complex special needs' (Department for Children, Schools and Families, 2008: 87). BESD is different from many other forms of Special Educational Need (SEN) in that it is socially defined, rather than related to a specific medicalised diagnosis:

> [a]n emotional or behavioural disorder is whatever a culture's chosen authority figures designate as intolerable. Typically, it is that which is perceived to threaten the stability, security, or values of that society. Defining an emotional or behavioural disorder is unavoidably subjective, at least in part.
>
> (Kauffman, 2001: 23, in Holt, 2010: 12)

Numbers of children with BESD (or 'Social, Emotional and Mental Health difficulties' as renamed and redefined in the 2014 Special Educational Needs and Disability Code of Practice) are large and increasing. For example, in 2012 it was the third largest type of primary SEN in state-funded primary schools (Department for Education, 2012). Children with BESD, who are educated in both mainstream and special schools, arguably pose a heightened problem for the inclusion agenda. As Jull (2008: 13) argues, 'educators remain confounded by how best to respond to students whose particular special educational need seems to justify punitive disciplinary action, including exclusion'.

Thus schools remain broadly exclusionary places for pupils with BESD; as Youdell (2006: 126) notes, 'by virtue of the designation, pupils with BESD are rendered outside the normative centre of subject, learner and subjecthood'. Their emotions, ways of forming and engaging in social relationships and behaviour can be seen to be 'out of place' within the school context; as Holt (2010: 148) notes, 'diagnosis is often instigated in schools by a child's inability to conform to classroom norms and expectations'. With emotions integral to these diagnoses, as well as the forms of emotional governance under discussion, the chapter serves to further the call for the more sustained and focused discussion of (non-normative) emotions in relation to disability (Holt, 2010). It also contributes to understandings of emotional geographies (Davidson et al., 2007) through the argument that contemporary strategies of emotional governance contribute to the reproduction of social difference (particularly along lines of disability) via the English educational institution.

The chapter draws on wider research that looked at children and young people's social relationships and (dis)ability (e.g. Holt et al., 2012, 2014; Bowlby et al., 2014; Lea et al., 2015). The research explored how everyday practices across school, home and leisure activities (re)produce disability as a valued or devalued identity, and how disability intersects with other forms of social difference (namely class, gender and ethnicity) across these practices. In-depth qualitative research took place across nine schools in three local authorities (LAs) in southern England, one secondary school, one primary school and one special school in each. The schools chosen had a diversity of specialist facilities, levels of inclusion and socio-economic composition. The chapter makes use of data obtained via participant observation of schools and leisure spaces and in-depth interviews with pupils and teachers in two of these schools. These observations were overt and the pupils had been given written and verbal information about the project and were able to opt out of being observed. Pupils were asked if they would like to take part in interviews (and associated semi-participatory research), and we obtained written consent from pupils and their parents/carers. Approximately 30 days of observation took place in each of the case-study schools, 4–5 adult interviews and 12 student interviews.

The two schools that appear in this chapter were a primary level (4–11) mainstream school in a coastal LA and a primary level (4–11) special school for children with BESD in an urban LA. The schools both had high numbers of children considered to have BESD (according to school census figures from the academic year 2008–9, 100% of the children at the special school were designated as having BESD as a primary need, and 10% were designated as BESD as a primary SEN in the mainstream school, compared with 1% for the wider coastal LA at primary level). The extracts drawn from research diaries and presented here relate to children who were designated as having BESD as a SEN.

The chapter proceeds by outlining the conceptual framework of governance, locates SEAL as a form of emotional governance and offers an outline of the current literature on the relation between SEAL and BESD. The chapter then outlines our methods and case studies, before turning to discussions of the mode of delivery of SEAL, and the kinds of relationship that children designated as having BESD have with SEAL as suggested by the empirical material. The chapter will then briefly conclude.

SEAL as a form of emotional governance

Governance is a broad and, according to Bevir (2010: 1), 'ubiquitous' term which has been understood and mobilised in a multitude of ways. Here, the term is framed through Foucault's writings on government, and more recent interpretations of them. Rather than a hierarchical relation in which those governing determine directly what those being governed do, governance relies on the 'self-governing capacities of individuals and collectives' (Sorenson and Triantafillou, 2009: 1). Subjects are positioned as being able to reflexively monitor and intervene in themselves in order to create the 'right kind' of self. This positions subjects as a 'potential resource' that can be 'activated' (ibid., 2009: 1) in the required way. The ideal subject of this regime of governance is self-aware and capable of choosing how to act and, at the same time, holds the capacity to act. To analyse governance therefore requires attention to be paid to 'those processes that try to shape, sculpt, mobilise and work through the choices, desires, aspirations, needs, wants and lifestyles of individuals and groups' (Dean, 1999: 12–13). Rather than simply being regimes of 'correcting and disciplining', relations to the self characterised by 'augmentation and improving' (Sorenson and Triantafillou,

2009: 3) are important here. State action indirectly attempts to circumscribe permissible actions and behaviours, against social and cultural norms of correct and incorrect conduct.

Such self-governing capacities are not seen to be innate characteristics, but instead rely on the production of institutional settings that foster the capacity to act on the self in this way. Indirect state action becomes dispersed via a variety of non-state institutions and citizen activities, and across a plurality of sites. These (often state fostered) institutional settings operate as 'part of a "capillary" of relations in which power continually circulates and re-circulates' (Barnes et al., 1998, in Newman, 2001: 21). Educational institutions offer an important example of this; creating subjects who are able to (and want to) self-govern has come to be a key purpose in education, reflecting the broader aim of creating future citizens (Staeheli, 2010). As part of a wider shift, in which the attention of policy makers has shifted from 'structures and processes' to 'personal skills and self-efficacy' (Gillies, 2011: 186), emotions have been positioned squarely at the centre of the capacity to self-govern (Rose, 1999). This has been understood as part of a broader desire to 'insinuate emotional competence into young subjectivities for the purpose of creating a more effective and governable future citizen' (Gagen, 2015: 147).

There has been a vast increase in UK school-based programmes focused on various interpretations of social and emotional wellbeing over the past 15 years (Watson et al., 2012: 57). SEAL is the most prominent of these, and by 2010 90% of primary schools and 70% of secondary schools were using it (Watson et al., 2012: 57). While the government's review of Personal, Social, Health and Economic education (Department for Education, 2013), under which SEAL lessons sit, means that it remains non-statutory and no new curriculum material will be produced, schools in England have continued to use the SEAL resources in a variety of ways. SEAL is designed to have impacts across all areas of school and is delivered via discrete lessons as well as being incorporated in subject-based teaching. Gagen (2015: 142) sees SEAL to have taken on such currency in schools that she argues 'emotional competency is becoming the new benchmark by which young people are marked as successful subjects', thus underlining the increasing significance of emotional governance as a process operation not only in schools in England but more broadly in society.

SEAL is based largely in the work of Daniel Goleman, whose popular psychology writings on emotional intelligence have been taken up in workplaces towards the development of more emotionally attuned and effective modes of management. Holmes (2010: 146) suggests that the incorporation of notions of emotional intelligence in the workplace also underpins a shift in understanding; previously, 'doing emotions "well"' was seen to be a form of socialised skill learnt tacitly, whereas Goleman's ideas reframe it as a matter of work which can be (and *should be*) explicitly taught and learnt. Goleman's writings (and the associated change in understanding around what it takes to do emotions 'well') underpin the five aspects of 'social and emotional learning' that constitute SEAL – self-awareness, managing feelings, motivation, empathy and social skills. In practice, while individual schools, and indeed classrooms, take SEAL on to differing degrees and in different ways, it can be seen to frame a shift towards schools developing and working with particular understandings of what emotions are and what emotional competency looks like, as well as the emotions being an explicit concern and responsibility of schools.

SEAL provides a framework that creates and formalises norms around how (individual) bodily feelings and sensations equate to particular emotions, and sets out what the proper response to such feelings and sensations is. This means that the children are encouraged to become reflexive subjects, and this reflexivity is routed through the lens of the emotional (Holmes, 2010). They turn their gaze inwards to become familiar with their bodily feelings, sensations and thoughts, and then compare these knowledges of the body to the definitions of different emotional states as laid out in the SEAL material. The SEAL material places the sensations that the children experience at the level of their bodies into wider socio-cultural frames of meaning, and in turn emotions become a lens through which children can name and make sense of the bodily states and feelings they experience. In SEAL, this way of knowing the self underpins the development of an ability to manage the self, to intervene in, control and modify bodily states (and thus emotions) using the strategies and techniques that are outlined in the SEAL curriculum. SEAL, therefore, labels personally experienced sensations and experiences as socially and culturally categorisable emotions, positions these emotions as *controllable*, and constructs children to be actors who are *able* to control them. The *right kind* of emotionally self-governing subject, as imagined via SEAL, is one who has *identifiable* feelings and who is *able* to intervene in their emotions and *manage* their moods. They can rise to the responsibility that SEAL places upon them to manage their emotional selves adequately.

What does SEAL do to children labelled as having BESD?

Many different lines of critique have emerged in relation to school-based forms of emotional governance, and also SEAL: for example the problematic science that is seen to underpin it (Claxton, 2005, in Gillies, 2011), the way the focus on the individual obscures wider, more structurally embedded, inequalities (Gillies, 2011), the problematic generalisations about what the subject is and what its emotional needs are (Ecclestone, 2010), and the way in which these produce and reproduce inequitable expressions of social difference (e.g. gender, class) (Ecclestone and Hayes, 2009), as well as the consequences of SEAL for pupils seen to have 'problematic' emotions. At first glance, such forms of emotional governance might be seen as an effective way of solving some of the problems that have come to define BESD as a diagnosis. However, the more established interpretation is that such forms of emotional governance are problematic for pupils who are designated as having BESD, in a number of ways.

First, SEAL problematises students with BESD by effectively establishing and formalising certain emotional or behavioural responses as 'normal'. Such norms are not, in themselves, inherently problematic, but when they come to limit possibilities by encouraging 'subjects to become highly efficient at performing a narrowly defined range of practices' (Taylor, 2009: 47), they become an issue. The division between normal and abnormal becomes culturally embedded, thus obscuring the fact that these norms are constructed (rather than natural and neces- sary). Within schools historically there have been many material consequences of this, particularly in the disablement and exclusion of children who do not meet norms across a range of mind–body differences (Holt, 2004, 2007, 2010). Along similar lines, Ecclestone (2010: 67–8) argues that the kinds of 'claims about the ease and efficacy of developing emotional skills' that are contained within SEAL 'conceal highly normative and complex moral judgements and decisions'.

Second, SEAL misrepresents and negatively constitutes some emotions (such as anger) that might (legitimately) be part of the identities of pupils with BESD. Ecclestone (2010: 63) notes that SEAL positions negative emotions as parti- cularly problematic. Gagen (2015: 147) notes that anger is seen to be the worst, being 'cast as a dangerous and explosive emotion' which has deleterious con- sequences for the current (and future) lives of pupils. She goes on to suggest that the programme response to anger is 'not understanding but eradication' towards a particular regime of emotional governance in the classroom and wider school, and an atmosphere characterised by a 'calm, emotionally flat ideal' (Gillies, 2011: 189) in which the 'actual display of emotions . . . marks out the emotionally illiterate' (Gillies, 2011: 187). This highly regulated approach to emotions reinforces the positioning of pupils designated as having BESD as somehow 'other' (Watson et al., 2012), as well as contributing to the construction of their identities as illegitimate within the school setting. Relatedly, Gillies (2011, in Gagen, 2015: 147) has argued that the emotional literacy agenda (of which SEAL is part) does not adequately deal with, and indeed is injurious to, the 'complex, fraught and socially embedded emotions that many students struggle with in their everyday lives'.

Third, SEAL can reproduce the representation of socio-emotional difference as an individual pathology. By locating the 'solution' within the child, wider social contexts and inequalities (which are generally understood to play a part in producing emotional and behavioural differences) are effectively hidden. The creation of the subject (as responsible for their own wellbeing) can obscure wider structural inequalities in society, and shift the causation and responsibility away from the persistently problematic structural inequalities (Gillies, 2011; Watson et al., 2012). If individuals don't respond to the kind of 'responsibilisation' that SEAL depends upon, they run the risk of being seen as not availing themselves of the opportunities provided by the state (Watson et al., 2012: 78) – not because they *can't* but because they *won't*. SEAL thereby can play a part in reproducing an understanding of children with BESD as the problem, rather than upholding the 'responsibilisation' of the subject as the potentially problematic element.

Spaces of exclusion for children labelled as having BESD

As outlined above, many of the writings on the relation between SEAL and BESD focus on the curriculum content. In order to augment this existing literature, this chapter approaches the relation between SEAL and BESD in slightly different ways. The first of these is via the question of the contribution and function of the 'sites where knowledge is spatialized' (Gagen, 2015: 142), in this case the classroom and, more specifically, circle time. Second, the chapter looks at the model of the self-reflexive subject that SEAL relies on and, via staff interviews, asks what this *right kind* of subject, as constructed via SEAL, is seen to do to pupils designated as having BESD.

Participation: circle time

Circles are a common mechanism for the delivery of SEAL and require the classroom to be rearranged so that the children and adults are sitting on their chairs in a circle. With its roots in restorative justice (see Lea et al., 2015), the idea of 'circle time' is that everyone has an equal opportunity to speak. Both schools under discussion here used circles as a means of delivering SEAL. Looking at circle time offers some insights into how SEAL lessons are delivered in the classroom – in other words, the kinds of institutional settings that are seen to foster the capacity of pupils to become the *right kinds* of self-governing subjects (who are able to develop a reflexive gaze, identify emotions, and act and manage those emotions) as outlined above. What follows are research diary extracts from SEAL circle time in both schools which show how children designated as having behavioural, social or emotional difficulties were variously excluded from participating in the SEAL curriculum. While most of the observations from the research diary involved other parts of the curriculum, in the mainstream school it was common for the children named here to be out of their seats, not engaged with the activities the teacher had planned, or even out of the classroom. In the special school, whilst the children in this class often found it difficult to engage in lessons where forms of structured participation were required (e.g. sitting in a certain way, following specific instructions), the particular instance set out here displays a higher degree of non-participation than was the norm.

The first extract is from the mainstream school classroom of year 2 children (normally aged between 6 and 7) – the children were in a circle while a SEAL lesson was being delivered about anger. They had been talking about skills to help them relax and to get rid of anger (all names are replaced by pseudonyms):

> The children were asked to work in 3's. Campbell, a child designated as BESD, got up and left the circle to get a drink of water (while this could be seen as him being thirsty, it could also be interpreted as him finding it a difficult topic, or being asked to talk in a three being unviable for him). On rejoining the circle, Campbell never really settled down. He started rolling his body around, touching the girl sitting next to him on the head. This made the teacher ask him if he needs some time out from the circle. He didn't respond to the teacher and began messing with his water bottle and shouting out. This time he is asked to leave the circle for a minute – he does this but goes and lies on a table. Another adult in the classroom asks him to go to the

corner . . . Danny, another boy designated as BESD, wasn't in the circle, but instead was unpacking one of the trays next to me at the back of the classroom. His individual needs assistant came across and repacked the tray, then took him off to the quiet corner and sat with him. After this, the other year 2 teacher took Danny and his INA [individual needs assistant] out of the class.

The second extract comes from the special school, in which all the children were designated with BESD as their primary or secondary form of SEN. The observations took place in the classroom with the oldest children in the school (aged between 9 and 11). An educational psychologist (EP) from the council had come in to do 'Our time' with the pupils, a curriculum intervention based on SEAL. The EP explains to the pupils that 'Our time' will let them talk about their feelings, learn about how other people think and feel, and talk about and practise what to do in difficult situations:

> Brandon, Lisa, Sienna and Adam are all standing up. Adam sits down back in the circle and the teaching assistant rubs his back. Adam pushes back from the circle and doesn't join in. The class teacher says that Adam needs to be back with us . . . Brandon has left the classroom, Theo is already out. Lisa runs out too. Adam tries to go to the door to look. It's just Jared, Jason, Adam and Sienna left. Brandon comes back in, Adam gets up and says 'let's go and join the bandwagon' and runs about the classroom.

In both of these extracts, the children were not in the circle all the time, and some of the children did not join the circle at all. While the mainstream class teacher told us that the children not in the circle are 'still . . . listening to it all and they may still know what we've been talking about', this was certainly not true for Danny who left the room halfway through circle time. Therefore, in order to access the SEAL curriculum and to sit in (and participate in) the circle, the children already need some degree of ability to manage their emotions. It seems, then, that these children were unable, or were differentially able, to access the SEAL curriculum because they could not, or did not, sit in the circle. As Holmes (2010: 146) points out, different people 'come to emotion work differently prepared' and this prerequisite privileges those children who come to school already equipped with, or who are able to quickly learn, the level of self-regulation required to be able to sit in a circle (let alone reflect on potentially difficult feelings and emotions).

As Holmes (2010: 146) argues, 'understanding social and self-reproduction can benefit from attention to emotions' and SEAL can be seen as a form of emotional governance that reproduces inequalities on the basis of different, non-normative, forms of emotions, socialities and/or behaviour. Being unable, or differentially able, to access the SEAL curriculum has consequences for the children. For example, circle time becomes another part of school that is disliked – indicative of this is that one of the children at the special school renamed 'Our time' 'Shit time'. Additionally, it means that the children do not access the SEAL curriculum, which has a number of potential consequences (differing across the settings under examination here). In the mainstream school, not accessing the SEAL curriculum negatively affects the identities of the children who are not included – these pupils are marked out as abnormal and their difference within the classroom is underlined and reproduced. The SEAL curriculum negatively represents (or misrepresents) the differences that are seen to characterise Campbell and Danny; the lesson that this extract came from was one concerned with the management and elimination of anger in the pupils. Furthermore, assuming that the children who stayed in the circle could access the SEAL curriculum, they would be equipped to move closer to the norm that SEAL circumscribes; this means that those who are not in the circle run the risk of being doubly disadvantaged as the classroom (and school) based norm of behaviour moves even further away from their differences, thus increasing the distance between the children designated as having BESD and their peers.

Something a bit different could be said to be happening in the special school. Here, the children remaining in the circle were in the minority (there were only eight children in the class, and out of those only three remained in the circle), and so this SEAL lesson can be seen to mark out those who stay in the circle in terms of difference; the line between normal and abnormal is differently drawn here, and characteristics of compliance and participation are marked out as abnormal. So in the classroom and wider school those who take part in SEAL are those who are constituted as different. If the special school is put in a wider context, however, it could be suggested that participation (or not) in SEAL (as a universal intervention operating across both special and mainstream schools) widens the gap between the mainstream and special school. This could create problems in the future; for example, when many of the children at this special school move back into mainstream schools for Secondary (11–18), or even when the pupils complete their compulsory schooling.

We asked about these observations in in-depth interviews with staff members (the mainstream school class teacher and the headteacher of the special school), and in their interpretations the staff offered us (contrasting) explanations for the children not being in the circle. The mainstream teacher suggested that Campbell and Danny 'opted out' of the circle, thus framing their lack of participation in terms of a choice. In contrast, the headteacher in the special school suggested that participating in circle time was not a matter of choice, but rather a question of ability or inability to sit in this way:

> Well almost circle time becomes a measure of the inability to interact, rather than a means to enable interaction . . . So if you can't sit in a circle, then circle time's going to be another barrier.

In suggesting that the child is choosing not to take part, the mainstream teacher effectively situates the cause of not accessing SEAL in the child (while not necessarily blaming them), whereas the special school headteacher said that the children *cannot* access the circle, thus shifting the causation to how the SEAL curriculum is delivered.

In the language of self-governance via the emotions, neither set of pupils is 'activated' to become the *right kind* of self-governing subjects that SEAL demands, but, by suggesting that the children in the mainstream school lack the *desire* to become self-governing (rather than being *unable* to become self-governing), these subjects are even further problematised. If in the present they are choosing not to participate in SEAL, at the same time they are not stepping up to the responsibility that the state inculcates in them to become emotionally literate citizens. As noted above, this runs the risk of framing the pupils as rejecting the opportunities provided to them by the state (Watson et al., 2012: 78) and locating the 'solution' within the child (they can choose to take part), so shifting attention away from the wider social contexts and structural inequalities that are seen to play a part in producing pupils who come to be understood via the designation of BESD. In contrast, locating the problem in the way the SEAL curriculum is delivered does some work to mitigate these problematic understandings.

Participation in SEAL in the mainstream school

These different interpretations of why the children were not in the circle in turn gave rise to different understandings of how they might be enabled to participate in the circle and, further, different understandings of the value of engaging with the SEAL curriculum. The mainstream teacher told us that repeated opportunities to return will enable the children to access the circle:

> I still feel that with circle time that eventually it is accessible by all of them but the key thing is to keep giving them an option to come back in.

The emotional norms and forms of self-governance entailed in SEAL were not questioned by the mainstream teacher, and were assumed to be attainable by all. The teacher was straightforwardly accepting of the benefits of SEAL: 'Well I think here we spend so much of our school time dealing with children with huge emotional issues that what SEAL did was to give us a bit of a structure to do it within.' She went on to describe the value of SEAL for the children in the school:

> it gives them the words to talk about it . . . to start really focusing on what happens to your body when you're feeling those things (so they can learn to start recognising – oh I'm not breathing, I feel really hot I think I'm really angry). Right, what are those things we've talked about to stop this thing? And I think that's really powerful . . . for children who have a lot of extreme emotions.

In both of these quotes, the teacher explicitly outlines the value of SEAL for the children, such as Campbell or Danny, who have 'extreme emotions' and 'huge emotional issues'. If these children were participating in the circles or even in the room during all of circle time, then these statements by the teacher could be taken at face value, but it seems that the empirical observations problematise the suggestion that SEAL has value for pupils designated as having BESD.

Participation in SEAL in the special school: a different view

In contrast, the special school headteacher outlined a different understanding of how children might be enabled to participate and, in particular, offered an interpretation that critiqued the value of SEAL for pupils designated as having BESD:

> So how are you going to enable this child to sit in the circle? My view would be it's about building trusting relationships ... becoming more resilient, being able to take a few risks, and they ultimately lead into a point where children will do stuff that they don't want to do. And that's where circle time comes in because most of the children don't really enjoy circle time because it's a bit of a spot light on their frailties, on their sort of vulnerabilities! Why would they want to do that?!

The headteacher effectively says he cannot blame the pupils for not wanting to be in the circle, because the kind of reflexive gaze that SEAL is based upon and which is necessitated in the circle is too difficult for the pupils to contemplate. The subtext in the headteacher's interview is that the EP leading the lesson is a less trusted 'outsider' who has not had the opportunities to build trusting relationships with the children over time (in contrast with the usual teacher). The reflexive gaze and the outsider leading the lesson combine to make SEAL *less accessible* to the students than the usual curriculum.

Even if the pupils make it into the circle and are able to look inwardly to identify their sensations and feelings, the headteacher considers the pupils to have limited capacities to act on themselves:

> So in terms of presenting them with ... 'so this is how you're meant to behave in certain situations' [it] is useful, but the way it does ... fall apart is the pupils' ability to – it's not an active choice ... that they behave like that, it's a real drive to behave like that ... therefore – what's driving them to behave like that? That's what you address.

Here the headteacher suggests that the pupils don't become the kind of 'responsibilised' self that SEAL (as a form of emotional governance) requires not because of their own active choices but rather because of the wider structural factors (such as poor housing, lack of nurture) that drive the pupils' behaviour. In directing attention towards these wider structural factors, the headteacher highlights how SEAL frames the emotions in ways that reproduce inequalities (on the grounds of behavioural, emotional and/or social differences).

The headteacher questioned SEAL (and other such programmes imposed from outside the school), echoing Gillies' (2011: 196) argument that SEAL misinterprets or hides the 'complex social and emotional interactions characterizing pupils' lives and experiences':

> I'm sort of not overly sold on those external strategies . . . because . . . you go through all the process . . . for example . . . what was it we did last term, I can't even remember, sort of 'being good to be you' type thing – and yeah, you do all that stuff . . . half an hour later everything's fallen apart and . . . quite clearly you see it's not 'good to be you', is it?!

This comment about the use of external strategies highlights how programmes such as SEAL are problematically being applied in a prescriptive fashion irrespective of context (Boddy, 2013: 273). In this case it is perhaps because the EP from the council delivers the intervention. The school's everyday response to anger and other forms of emotional expression was based in an in-depth knowledge of the children that attended the school. It differed from SEAL's aim to remove or otherwise flatten out 'problematic' emotions and instead valued the expression of whatever the child needed to express, seeing the emotions as an important and constitutive part of them and understanding them as part of the (often difficult) wider social contexts and environments within which the pupils lived their lives outside school.

This explicit approach was made possible by the fact that the school was a special school with a focus on BESD. This made it possible for it to offer multiple forms of specialised response (often located in therapeutic grounds), such as having an art therapist resident on the staff, responses often inaccessible or less well developed in mainstream contexts. The existence of special schools is contentious; the Salamanca Statement, published in 1994 by UNESCO, argued that *all* young people with special educational needs should be educated within mainstream schools. This argument is made on educational and societal grounds (Morris, 1991; Holt, 2004, 2007; Thomas, 1997), and Holt argues that this division between special and mainstream education is seen to be 'pivotal to the negative reproduction of disability in society' (Holt, 2007: 786).

This special school can be interpreted as a space which is attempting to do justice to the complex and difficult emotions that often form the basis of the designation of BESD. However, the co-existence of such complex/difficult emotions and forms of emotional governance such as SEAL (which is often irrelevant to the children and serves to reproduce inequalities on the basis of different, non-normative, forms of emotions, socialities and/or behaviour) is contradictory. The ongoing use of SEAL within the school (despite the quite vehement opinions of staff that it doesn't work for these particular children) demonstrates the persistence and prevalence of this model of emotional governance in the English educational institution. At the same time, the model of the 'responsibilised' emotional subject demanded by SEAL persists. While it is challenged within the school, this is not communicated outside of the school. As a result the work done to value the identities of children designated as having BESD, and to reframe the relationship between those children and participation in SEAL away from a language of choice, travels no further than the school itself. While the school internally values the differences that are seen to be characteristic of the designation of BESD, its existence effectively contributes to the reproduction of the broadly negative understanding of children with BESD in mainstream schools and wider society.

Conclusions

This chapter has sought to begin to explore the complex relationships between SEAL (as a form of emotional governance) and forms of emotional expression that have come to be problematised via categories such as BESD. Generally, SEAL is seen to contribute to and consolidate the exclusion of children who are considered to have behavioural, emotional and social differences at a variety of scales, both within schools and also within wider society. The case-study schools suggest that this happens across a range of settings, including mainstream and special schools, and to children of different ages. The chapter has built on the existing literature that has addressed the relation between SEAL and BESD via its focus on the spatialisation of the types of knowledge that form the basis of SEAL, and has offered a consideration of how specific classroom spaces variously contribute to the exclusion of children with BESD from SEAL curriculums, or reconfigure the imperative to participate in SEAL.

Ecclestone (2007: 464) argues that SEAL is situated within 'wider cultural narratives in which everyone is seen as emotionally vulnerable', which suggest that such forms of emotional malaise and difficulty can be diagnosed and remedied with the right kinds of professional help. Indeed, SEAL and other such forms of emotional governance were originally developed as a response to concerns about the emotional lives of such vulnerable groups as children with BESD (Ecclestone, 2010) and worries over classroom behaviour and discipline (Francis and Skelton, 2006, in Ecclestone, 2007). Somewhat paradoxically, however, it seems that this form of emotional governance has contributed to the production of such pupils as *too* emotionally difficult to participate in SEAL, across various classroom settings. Furthermore, if Gagen (2015: 147) is right in her argument that teaching forms of emotional governance (such as SEAL) begins to 'shape emotionality as a vital feature of future civil life', the production of pupils seen to have BESD as potentially *too* problematic to participate in SEAL might well create further and future forms of exclusion. If new limits to citizenship (based on the ability to develop a relation of self-governance of the emotions) are set, the potential impacts of such changes on pupils designated as having BESD are ones which remain to be addressed in any systematic or coherent way.

Many thanks to the schools and individuals that took part in this research. The research was funded by the Economic and Social Research Council (RES-062-23-1073-A).

References

Bevir, M. (2010) 'Governance as theory, practice, and dilemma', in Bevir, M. (ed.) *The Sage handbook of governance*. Sage, London, pp. 1–16.

Boddy, J. (2013) 'European perspectives on parenting and family support', in Ribbens McCarthy, J., Hooper, C.-A. and Gillies, V. (eds) *Family troubles? Exploring changes and challenges in the family lives of children and young people*. Policy Press, Bristol, pp. 263–78.

Bowlby, S., Lea, J. and Holt, L. (2014) 'Learning how to behave in school: a study of the experiences of children and young people with socio-emotional differences', in Mills, S. and Kraftl, P. (eds) *Informal education and children's everyday lives: geographies, histories and practices*. Palgrave Macmillan, Basingstoke, pp. 124–39.

Davidson, J., Bondi, L. and Smith, M. (2007) *Emotional geographies*. Ashgate, Farnham.

Dean, M. (1999) *Governmentality: power and rule in modern society*. Sage, London.

Department for Children, Schools and Families (2008) *The education of children and young people with behavioural, emotional and social difficulties as a special educational need*. DCSF, London.

Department for Education (2012) *Statistical first release: special educational needs in England, January 2012*. Department for Education, London.

Department for Education (2013) *Guidance: personal, social, health and economic education (PSHE)*. Department for Education, London.

Ecclestone, K. (2007) 'Resisting images of the "diminished self": the implications of emotional well-being and emotional engagement in educational policy', *Journal of Education Policy*, 22(4): 455–70.

Ecclestone, K. (2010) 'Promoting emotionally vulnerable subjects: the educational implications of an "epistemology of the emotions"', *Journal of the Pacific Circle Consortium for Education*, 22(1): 57–76.

Ecclestone, K. and Hayes, D. (2009) *The dangerous rise of therapeutic education*. Routledge, London.

Foucault, M. (1991) 'Governmentality', in Burchell, G., Gordon, C. and Miller, P. (eds) *The Foucault effect: studies in governmentality*. University of Chicago Press, Chicago, IL, pp. 87–104.

Gagen, E.A. (2015) 'Governing emotions: citizenship, neuroscience and the education of youth', *Transactions of the Institute of British Geographers*, 40(1): 140–52.

Gillies, V. (2011) 'Social and emotional pedagogies: critiquing the new orthodoxy of emotion in classroom and behaviour management', *British Journal of Sociology of Education*, 32(2): 185–202.

Holmes, M. (2010) 'The emotionalization of reflexivity', *Sociology*, 44(1): 139–54.

Holt, L. (2004) 'Childhood disability and ability: (dis)ableist geographies of mainstream primary schools', *Disability Studies Quarterly*, 24(3), at: http://dsq-sds.org/article/view/506/683 (accessed 22 June 2016).

Holt, L. (2007) 'Children's sociospatial (re)production of disability within primary school playgrounds', *Environment and Planning D*, 25(5), 783–802.

Holt, L. (2010) 'Young people with socio-emotional differences: theorising disability and destabilising socio-emotional norms', in Chouinard, V., Wilton, R. and Hall, E. (eds) *Towards enabling geographies: 'disabled' bodies and minds in society and space*. Ashgate, Farnham, pp. 145–65.

Holt, L., Lea, J. and Bowlby, S. (2012) 'Special units for young people on the Autistic Spectrum in mainstream schools: sites of normalisation, abnormalisation, inclusion and exclusion', *Environment and Planning A*, 44(9), 2191–206.

Holt, L. Bowlby, S. and Lea, J. (2014) 'Emotions and the habitus: young people with socio-emotional differences (re)producing social, emotional and cultural capital in family and leisure space-times', *Emotion, Space and Society*, 9: 33–41.

Jull, S. (2008) 'Emotional and behavioural difficulties (EBD): the special educational need justifying exclusion', *Journal of Research in Special Educational Needs*, 8: 13–18.

Lea, J., Holt, L. and Bowlby, S. (2015) 'Talking about socio-emotional differences, reproducing "Behavioural, Emotional and Social Difficulties"? The use of restorative approaches to justice in schools', in Blazek, M. and Kraftl, P. (eds) *Children's emotions in policy and practice: mapping and making spaces of childhood.* Palgrave, Basingstoke.

Morris, J. (1991) *Pride against prejudice.* Women's Press, London.

Newman, J. (2001) *Modernising governance: New Labour, policy and society.* Sage, London.

Rose, N. (1999) *Governing the soul: the shaping of the private self.* Free Association Books, London.

Sorenson, E. and Triantafillou, P. (2009) 'The politics of self-governance: an introduction', in Sorenson, E. and Triantafillou, P. (eds) *The politics of self-governance.* Ashgate, Farnham, pp. 1–24.

Staeheli, L. (2010) 'Political geography: where's citizenship?', *Progress in Human Geography*, 35: 393–400.

Taylor, D. (2009) 'Normativity and normalization', *Foucault Studies*, 7: 45–63.

Thomas, G. (1997) 'Inclusive schools for an inclusive society', *British Journal of Special Education*, 24: 103–7.

UNESCO (United Nations Educational, Scientific and Cultural Organization) (1994) *The Salamanca statement and framework for action on special needs education.* UNESCO, Paris.

Watson, D., Emery, C., Bayliss, P., Boushel, M. and McInnes, K. (2012) *Children's social and emotional wellbeing in schools.* Policy Press, Bristol.

Youdell, D. (2006) *Impossible bodies, impossible selves: exclusions and student subjectivities.* Springer, London.

8 'Supporting People'

Regulation, welfare practice and emotions

Rachael Dobson

This chapter uses policy and practice responses to vulnerable people to think about the conceptual relationship between the emotions and governance in a marketised, modernised, networked and mixed-economy welfare state (Lewis, 2005; DTLGR, 2001; Clarke et al., 2000; Clarke and Newman, 1997). It tracks what happens when a voluntary sector supported-housing organisation is brought *into* the state and becomes 'governable terrain' via mechanistic and fiscal processes (Carmel and Harlock, 2008: 156). The analysis developed here draws on empirical qualitative research with five housing and homelessness organisations in a metropolitan area in the north of England, UK, in 2008–9, which involved observation and interviews with 30 practitioners in front-line and managerial roles. It focuses on findings from a voluntary sector supported-housing organis- ation, anonymised here as the Windmill Project. Established in the late 1970s, the Windmill Project provided services for homeless and vulnerable in the local area, ranging from 'floating support' delivered to clients in their own homes, to more traditional hostel-style and staff-supervised accommodation where clients had their own room in a house with shared kitchen, bathroom and living room facilities.

In the UK context, statutory rights to housing are enshrined in homelessness and housing law, but *support* services have traditionally been run by a mixed economy of statutory, voluntary, community, charitable and faith-based organis- ations (Johnsen, 2014; Dobson, 2011). The Windmill Project was, at the time of the research, contracted and majority-funded by the local authority to provide tem- porary accommodation and resettlement services for homeless people through the ring-fenced 'Supporting People' programme (hereinafter Supporting People), which was introduced in 2003 under the New Labour government (Cooper, 2013; ODPM, 2004). That funding programme is now much altered (Bury, 2011). In 2009, under the Conservative–Liberal coalition, the government removed the ring-fence from Supporting People, meaning that local authorities were free to spend the money elsewhere, resulting in the loss of service-level agreements and funding for voluntary sector housing services (Butler, 2014).

While it will be important to think about the effects of those changes, the focus of the present chapter remains on the Windmill Project. This is because tracking developments under Supporting People provides a strong case study for the conceptual analysis of a relationship between the emotions and governance. Specifically, the advent of Supporting People triggered an altered relationship between the state and voluntary sector organisations because of regulatory and audit processes implemented by the local authority. Empirical findings showed that one specific target stood out for Windmill Project workers. They were required to 'move on' a client from temporary accommodation within six months as opposed to a former organisational target of two years.

Within social policy and welfare debates, there is a tendency to think about matters of regulation, audit and targets as undermining obstacles to workers' efforts to exercise and develop their professional knowledge, and to get on with their jobs. It is claimed, for example, that 'professionalised' welfare work has become *more* about the ability to flourish within audit regimes via administrative and technical market-oriented accountability and 'managerial' capacities (Carmel and Harlock, 2008: 164) and *less* about a type of public service ethos, values or care (Hoggett et al., 2009: 159; Miller et al., 2006). Relatedly, audit cultures stand accused of creating tick-box cultures and 'games' whereby workers narrate an experience of practice via audit documentation that bears limited resemblance to workers' realities, but it is done to develop the *impression* of accountability and 'get by' within audit regimes (Buckingham, 2012; Lipsky, 2010: 163; Hoggett et al., 2009: 155; Milbourne, 2009; McGivern and Ferlie, 2007: 1371). Research with the Windmill Project resonated with those arguments. However, the findings presented here also offer alternative contributions to those debates.

Windmill Project practitioners did not reject regulation and audit processes out of hand, and nor did they dismiss the notion that they *should* work towards a 'move-on', but they did contest the *nature* of responses to homelessness and vulnerability under Supporting People. The shift to a six-month move-on was thought to result in shallower and more superficial levels of support because it did not enable a type of *endogenous* change in the short or long term. Workers thought that the funding regime, and the regulatory processes that it engendered, could lead to *absences* of care and professional *failings* by delimiting the sorts of work that could potentially lead on to endogenous change, rendering support workers somewhat ineffectual. There was also evidence that Supporting People's strategic ambitions and associated targets were at odds with practitioners' sense of a spatial, affective, social and psychic *distance* between themselves and their clients.

In summary, the 'move-on' target was the subject of intense interest and concern for Windmill Project workers because it stimulated tensions between what practitioners thought constituted support, on the one hand, and how this was operationally achievable in light of the governance of homelessness services under Supporting People, on the other. In contesting Supporting People, practitioners evoked clients' emotions to question contemporary forms of service provision, conceptualised in this chapter as neoliberalising: marketised, networked and regulated. This encompassed client reactions to Windmill Project workers in interpersonal interactions, to more ontological explanations of what practitioners thought it meant to exist within and beyond institutional spaces as a vulnerable person. Those findings are unpicked in this chapter by drawing on relational, psychosocial and critical feminist informed social policy and welfare scholarship, which foregrounds the intellectual potential of the emotions for deepening analyses of human power, agency and subjectivity, for understanding both researcher and research participant, and for developing critical approaches to the social relations of the state.

Governance and the emotions

This chapter draws on three broad sets of claims about a conceptual relationship between governance and the emotions. First, the study of emotions recognises the effects of the author's *involvement* with the research as a practitioner in one of the participating organisations, the Homelessness Unit. As argued elsewhere (see Dobson, 2009), emotional reactions to data presented in this chapter were relationally structured by the researcher's immersion and institutional positioning in housing and homelessness practice. Those reactions have potential to provide substantive insight into the multiple voices of contemporary governance regimes. That reflexive and affective approach to data analysis is a persistent undercurrent to the present discussion.

Second, bringing the study of the emotions to questions of human power, agency and subjectivity enables a social and relational approach to qualitative scholarship. For psychosocial scholars it means that human actors are not just understood as the bearers of discourse whose actions and understandings of the world are shaped by allocated identifications formed via cognitive, conscious, historical and discursive processes (Hoggett, 2008: 70). Rather, human power and agency are always formulated through relationships with others (Hoggett, 2008; Hunter, 2003), possessing rational and irrational, reflexive and unreflexive, conscious and unconscious capacities and an internal and feeling subjectivity (Hoggett, 2008: 70; Hoggett, 2001). Those approaches to human power, agency and subjectivity are especially important for thinking through why Windmill Project workers thought that Supporting People was at odds with vulnerable people's experiences of institutional spaces and capacities for personal change.

Third, claims about welfare practitioners' emotional reactions to their institutional contexts and users of services are a long-standing feature of social policy and welfare practice debates insofar as they highlight processes of meaning-making within specific geographical, socio-political and cultural locations (see Lipsky, 2010). Those more 'mainstream' constructionist approaches are extended through psychosocial and feminist interventions that bring questions of power, agency and the emotions to the critical study of governance and the state.

Observations of practitioners' intensifications of feeling around specific issues, such as passions for welfare work, questions of social justice and ethical dilemmas, are used to highlight the complexities of modern welfare work and responses to vulnerability in apparently unstable globalised contexts riven with social threats (Hoggett et al., 2009). Critical approaches to the emotions also suggest that human power and agency are *constitutive* of institutional space (Lea, 2008). For example, data in this chapter shows that emotions are performatively mobilised by Windmill project workers as they relationally and retroactively construct and narrate their practice roles and organisational contexts. Extending those different insights, Shona Hunter's feminist psychosocial concept of 'relational politics' develops the study of emotions and governance by claiming that the *distribution of power* and emotion are intimately connected in governance:

> Indeed, the emotions are productive of power in the sense that they constitute part of the means by which the state comes to be, they are integral to its gendered and raced orderings and are in turn part of the means by which the state enacts gendered and raced power. They operate as connecting devices, bringing together multiple actors and objects into the reasonably temporarily coherent form we think of as the state.
>
> (Hunter, 2015: 22)

Hunter's central argument is that the state doesn't exist as a thing in itself, but rather comes into being through the everyday processes of relational contestation: everyday actions, investments and practices of the multiple and shifting range of people and other material and symbolic objects that make up the state (Hunter, 2015: 5). The emotions, human experience and subjectivity are central to those processes and practices, acting as a connective medium between the individual and the social, mobilising and creating possibilities for contestation, resistance and social change. This chapter provides space to bring Hunter's account of the social relations of the state to the study of Windmill Project practitioners' actions.

Regulatory justifications: creating transitions and transformations

Housing support services were built up rapidly in the UK under the Labour government's 'Blair' years (Butler, 2014). Supporting People was a ring-fenced central source of funding for vulnerable adults who needed housing-related support (as opposed to domestic, personal or nursing care) that brought together seven housing-related funding streams from across central government into a single programme funded by the then Department for Communities and Local Government (Homeless Link, 2013). A year after its inception, the programme was funded at £1.8 billion (up from £1.4 billion) with 6,000 service providers and approximately 37,000 individual contracts to respond to vulnerable people (House of Commons, 2004: 3). The programme understood 'vulnerability' as people who: had been homeless or rough sleepers, were ex-offenders and at risk of offending and imprisonment, had a physical or sensory disability, were at risk of domestic violence, had alcohol and drug problems, were travellers, teenage parents, homeless families with support needs, elderly people and young people at risk, had HIV and AIDS, and had learning difficulties (ODPM, 2004: 2).

Supporting People was justified on the basis that funding streams were administratively complicated, uncoordinated, overlapping and lacking in strategic coordination across government departments (Jarrett, 2012). This was claimed to result in two problems: a failure to consider questions of value for money and quality of service provision, and a lack of responsibility for ensuring adequacy of support for vulnerable people. The latter problem was attributed to a failure to develop an appropriate range of *preventative* and responsive measures for welfare user needs. In the context of homelessness services, that failure meant that welfare users' vulnerabilities were dealt with as they emerged and/or users were directed to more 'acute' and high-support need services than needed because it was all that was available (Jarrett, 2012: 3).

Supporting People can therefore be understood as a strategic response to two concerns about a failure to help vulnerable people. First, there was not enough early and responsive intervention in order to identify and respond to people's needs before they worsened. Second, the type of support available was a particularly 'heavy' institutional response to vulnerability. As a result, the programme's aspirations for homeless people were structured through ideas about how social care, housing and homelessness institutions could both create and worsen their vulnerabilities. It was no longer acceptable (if it ever was) to allow people with vulnerabilities to remain in an institutional setting on a long-term basis, whether a hostel or a hospital. Instead, interventions were required to reduce the potential for, and risk of relapse into, homelessness.

The systemic and operational implementation of Supporting People's funding and regulatory framework was central to the programme's strategic ambitions to provide a range of more preventative and appropriate responses to vulnerability. Funded organisations were monitored via a devolved allocation body made up of partnerships of social services and housing authorities, probation boards and primary care trusts (based on unitary or county council boundaries), and administering authorities (top level local authorities) who implemented their decisions (Jarrett, 2012: 26). Those teams had a remit to identify gaps in local provision for housing-related need and competitively tender for and regulate services through contractual processes and fixed-term service agreements.

Interestingly, Supporting People's strategic vision, when coupled with local authority commissioning teams' activities, became relevant for *all* homelessness organisations in a local area, even if they were not Supporting People funded. This was because decisions about funding one organisation rested to some extent on assumptions about what others would provide, and what type of need and service user they would respond to. The result of that approach is interpreted here as a *transitional-tier* operational model, and this represented a distinctive vision of support. To clarify that model, it is useful to draw on the research that informs this chapter and its participating organisations: a social landlord organisation, a homelessness unit (both statutory providers), a drop-in centre and night shelter, a resource centre and the Windmill Project (faith-based charity, community and voluntary sector organisations respectively).

Of those organisations, just the Windmill Project and three beds in the drop-in centre's 12-bed night shelter were Supporting People funded. However, from a strategic-operational perspective, the funding rationale extended beyond those two organisations. For example, a client using the drop-in centre or attending the resource centre could be 'signposted' by practitioners to the homelessness unit for a statutory assessment of need under housing law (DCLG, 2006b). If the homelessness unit assessed a customer as entitled to statutory provision as a result of their homelessness and vulnerability, they could be referred to the Windmill Project for a temporary stay before 'moving on' into an independent social landlord tenancy. In other words, each organisation was imagined by the local authority to respond to different aspects of vulnerable people's needs at different times. Each organisation would 'work with' a service user so that they could move on to a different 'stage' in the operational model so as to *transition* into independent living.

In the context of the Windmill Project, 'working with' clients and the formation of a support plan were key to creating a move-on, and they were also sites of regulation and audit. The support plan was a written document formulated through a series of regular arranged meetings between Windmill Project 'key worker' and client. Support planning activities were outlined and made available to the local authority for regular monitoring through Supporting People workbooks. Where targets were breached and outcomes not achieved, organisations had to provide written commentary and explanation to the local authority. If the Windmill Project deemed that a client was not ready for a 'move-on' into the next tier of provision, a recordable outcome was still necessary for monitoring purposes. For instance, if a practitioner thought that a client required continued support, they could argue to 'keep' that client or they could work towards a referral to an organisation funded to offer longer-term residential support.

The process flows outlined above represent an ideal-type approach by assuming a linearity of human experience. That linearity is at odds with the research into the realities of homeless people's lives. Welfare users may enter and exit services at different stages (Johnsen and Teixeira, 2010), 'recovery' may not be achievable within the confines and contexts of designated services, and there is a broader socio-political and cultural climate that structures homeless and vulnerably housed people's lives (Robinson, 2011; Scanlon and Adlam, 2008). Existing empirical research into Supporting People has already covered constraints to successful resettlement. For example, Harding and Willett (2008) show that service managers and key workers are clear about the financial, social and emotional obstacles that clients face. These include the effects of prior institutionalisation, experiences of isolation in independent accommodation and intolerance from private landlords towards homeless people (Harding and Willett, 2008: 436).

Notwithstanding those findings, this chapter maintains focus on Supporting People's normative approach because the *transitional* model observed so far complements and connects to *transformational* policy languages that were distinctive to the programme, *and* because those transitional and transformational narratives were precisely the subject of contestation and tension for Windmill Project workers.

While methodologically partial (Newman, 2000), the analysis of housing and homelessness policy languages under the UK Labour government between 1997 and 2010 highlights a set of distinctive 'messages' about how to best respond to vulnerable people (Dobson and McNeill, 2011). The emergence of early and preventative interventions and movement towards a transformational vision of support can be observed, for example, through reconfigured conceptions of long-standing support actions and activities. Supported housing would not *just* provide temporary accommodation. Rather, rough sleepers would be 'brought in from the cold' (Rough Sleepers Unit, 2000; ODPM and SEU, 2004), 'no-one would be left out' (DCLG, 2008) and hostels could become 'places of change' (DCLG, 2008; DCLG, 2006a, 2006c; ODPM, 2005). 'Practical' advice was now discussed in more proactive terms as 'solution' focused, with a consumer-like emphasis on showing welfare users their 'options'. Medical and counselling interventions now included reference to treatment through cognitive-behavioural therapies. Voluntary work was understood as a meaningful activity in ways that have contributed to a reconfiguration of voluntary sector organisations as social enterprises as opposed to charities and the idea that responsibilisation was key for evoking personal change in clients (Whiteford, 2010).

In summary, the foundational drivers for the Supporting People programme and its operational implementation highlighted a transitional vision of support. That vision dovetailed with a broader homelessness policy climate that understood vulnerable people as transformable via the 'right' type of proactive and positive homelessness intervention, organisation and practitioner. The argument developed in this chapter is that this approach implicitly (and occasionally explicitly) positioned vulnerable people as knowable, diagnosable and changeable. The next section shows how that policy climate stood in tension with practitioners' views.

Tensions in practice

Windmill Project practitioners' responses to the move-on target highlighted distinctive tensions for practice. Specifically, Supporting People's visions of support and the governance of homelessness organisations were deemed incompatible with the ontological texture of clients' lives. This section explores those points by way of reference to responses from three Windmill Project workers: Faye, Glen and John.

For service manager Faye there was initially a pragmatic-physiological quality to her concerns about the move-on target. For example, she talked about how vulnerabilities, like mental ill-health issues, were episodic, and a client who seemed entirely ready to leave supported housing on one day could relapse the next. In that situation Faye thought that enforcing a move-on would worsen a client's mental ill-health problem, explaining that it would instead be better for the client to remain in supported housing to ensure stability and thereby avoid worsening vulnerability. At the same time, Faye also engaged with more ontologically driven conceptions of both stability and meaningful social and emotional personal change for vulnerable clients through her experiences with younger clients:

> We have [young] people … that have come to us from *very* difficult backgrounds, and they've gone through university. It's not for everyone, it is

the odd ones . . . That has been achieved mostly when we've been able to keep them for quite a long time, because all we've done is normalise and stabilise their lives, and provided them sufficient of a springboard to get off. They've gone off into independence, it's taken longer than those two years or the six months, but they've sprung off into independence that the *world* recognises . . . They're back in the stream, aren't they?

(Faye, Service Manager)

Here Faye appears to move beyond diagnostic-medical explanations of personal change by engaging with questions about *why* stability (remaining in one space and place) is important. Faye's thoughts here are not unproblematic. In citing the perceived benefits of containment-as-care (Hunt, 1981), her views nourish the foundational principles of Supporting People, which regarded move-on out of an institutional setting as central to progressive client support. However, Faye identifies practitioners' tasks as less about diagnosis and treatment, and more about helping a client to reach a point of stable existence within their social worlds, perceiving this to have potentially positive effects that may be unknown for both practitioner and client in the moment. This is a more open-ended process that is claimed to offer a type of immediate *and* longer-term social and emotional autonomy to clients. For Faye, the contemporary push for a move-on could limit the sort of work that could potentially lead onto a type of more profound and meaningful *endogenous* change in clients' lives that spanned space and time (e.g. moving on from supported housing into higher education potentially a few years later). Faye thereby complexifies what constitutes a 'move-on' by raising questions about what it is and how it might be achieved.

Extending that finding, Windmill Project workers outlined two effects of Supporting People for clients' ability to achieve personal change. One effect suggested that the regulatory context for practitioner–client interactions led to client *disengagement* from support services, and another effect extended the aforementioned idea that the move-on imperative and governance of welfare provision in housing, homelessness and allied fields *created* and *worsened* clients' vulnerability. Beginning with the first effect, workers' responses raised questions about their perceived effectiveness in their interactions with clients. For example, Glen described the limits of a key worker session where he attempts to complete a support plan with a client in order to create a move-on:

I can sit here and tell you, 'if you do this, you'll get that, and if you do so and so, you'll get this'. To you, it's just pie in the sky really. You've never had it, you don't know how to go about it. Members of your family and your close friends haven't got it either. So why are you going to be so different? You're not. So I [as a worker, am] talking a load of rubbish really. So 'I'll just let you carry on talking and then when you're finished I'll go back, on my merry way, and go out for a pint or a smoke or whatever' [mimics service user] . . . They're going through the motions. They know what the game is . . . 'If I want to keep this roof over my head, I have to meet with this key worker, so I guess I'll go along and I'll do that'.

(Glen, Team Leader)

Glen describes how clients would not always comply with the type of support associated with creating a move-on: the key worker session. On face value, Glen presents the scenario as a straightforward mismatch between welfare users' lives, on the one hand, and Supporting People-driven organisational requirements, on the other. The client's reactions are presented as natural responses in light of Glen's perceptions about personal biography ('You've never had it, you don't know how to go about it. Members of your family and your close friends haven't got it either'). The client is afforded a conscious-reflexive capacity when Glen says what he supposes the client is thinking ('So why are you going to be so different?').

In talking about 'the game' that the client is engaged in, Glen's thoughts resonate with American anthropologist Robert Desjarlais's (1999) analysis of interactions between homelessness practitioners and the users of services. Desjarlais troubles institutional and worker tendencies to classify homeless people in terms of their 'needs', rather than understanding clients' performances of personhood as constitutive of institutional space and as embedded in human relations (Scanlon and Adlam, 2008; Lea, 2008). Bringing Desjarlais's work to the present discussion, there is a *performative* quality to the power and agency of Glen's client who enacts a particular subject position by demonstrating compliance *in order to* continue with a life that makes sense to him ('I'll just let you carry on talking and then when you're finished I'll go back, on my merry way . . .'). This is a shallow compliance insofar as it renders the key worker session superficial and Glen somewhat ineffectual, according to the terms set by the Supporting People regulatory process itself.

There is also a spatial and temporal quality to Glen's perspective insofar as he positions himself, and the Windmill Project, as peripheral to the client's life. In the interview he does not berate himself for *not* being the transformative professional or chastise the client over his *lack of* engagement with the support-planning process. He is not cynical about the client, and nor does he adopt a cultural-relativist position ('the client is in this position because it's all that he knows'). Rather, his comments suggest that he thinks the institution (yhe Windmill Project, Supporting People) is misplaced in thinking about itself as central to clients' lives, as having the regulatory power to create personal change in the face of a life that is embedded in historical and contemporary 'layers' of human, personal, social and institutional relations (see Hunter, 2015, p. 13). In another interview, John is also critical of what the institution imagines that it can achieve given the regulatory and systemic context. He explains:

You have clients, like a sausage machine, *spinning* through our service . . . In a few months with us, apparently you're supposed to turn their lives around. Between having a crisis and having full independence that gap is *huge* . . . my personal values used to believe that we worked with the issues that were barriers to secure housing, or the issues that provoked and promoted homelessness . . . unemployment, mental ill health . . . I don't think anybody's working in depth: the psychiatrist gets a half an hour, the community mental health worker gets three quarters of an hour, I get an hour a week with a client. We're all getting little bits of this client . . . so people just get actually crazier and more lost . . . we're putting people into empty flats expecting them to survive, people who we haven't actually had enough time to build up their personal resources to survive . . . in an empty flat . . . by ticking boxes as [service users] move through us do we *really* think we're affecting their lives? I don't think so. We're getting them a house but there might be other procedural and technical ways of getting a house without support workers. That might come next, support workers are doing so *fucking* little, they're not needed . . . We know how the system works, we can get the application forms, the bidding . . . we can pull in medical assessments, we can get them signposted to [organisations]. We can just speed it up a bit; but we're not doing actually anything to any depth.

(John, Project Worker)

Here, John is angry about a loss of 'traditional' voluntary sector welfare ethos ('my personal values used to believe'). He explains his view that professional welfarist judgement is devalued in contemporary housing-support work in favour of administrative abilities ('there might be other procedural and technical ways of getting a house without support workers'). The modernised and commissioning state is constructed as fragmenting and compartmentalising client needs. It takes on a dystopian quality in terms of its perceived effects for clients: 'spinning', 'crazier', 'lost'. The idea of 'tick-box' cultures is used by John *less* to think about a pretence or masking of what happens in practice, and *more* to express the idea that clients are disinterestedly and hyperactively processed through different organisations, which both represents and results in damaging refusals to recognise the complexity of human experience and subjectivity.

Conclusions

Existing social policy and welfare debates about regulatory and audit cultures argue that regulating welfare practitioners strips them of their creative autonomy and creates risk-averse environments and 'tick-box' cultures that do little to capture the realities of day-to-day welfare work, and act as undermining obstacles to professional practice. While findings in this chapter support those arguments to some extent, it is claimed that practitioners did not oppose the notion of targets and regulation out of hand. Rather, they protested more against a particular climate of interventions, contemporaneous visions of support and the governance of homelessness provision under Supporting People.

This chapter has claimed that Supporting People was part of a policy climate of transitional and transformative visions of how to best respond to vulnerable people. Empirical research findings have shown that those visions stood in tension with supported-housing workers' sense of what would meaningfully help service users. Those tensions were most strongly articulated through the perceived effects of Supporting People for the governance of local homelessness provision, and the service-level agreement that the Windmill Project had with the local authority, which meant that they had to move on clients within a six-month timeframe. This target placed responsibilities for the recovery of the users of services with the Windmill Project and its workers, and it represented a set of ideas about how these responsibilities could be achieved.

Tensions arose because Windmill Project practitioners did not regard clients as straightforwardly changeable. Rather, they understood clients as complex human subjects embedded in human, personal, social and institutional relations. From that starting point, acknowledging a complicated and relationally driven understanding of human subjectivity was seen to better enable a type of meaningful endogenous change. Practitioners observed a physical, social, emotional and psychic distance between themselves and the users of services. They could only provide a suitable *context* for personal change: they were not the *creators* of that change in any immediate sense, they were not privy to clients' interior subjectivity, and there was a more expansive social world structuring clients' experiences. In short, the move-on target mattered so much to Windmill Project workers because it disrupted how they understood the users of services. More instrumentally it mattered because it took away a vital ingredient for engendering a *context* for understanding clients and the potential for ontological change: time. Overall, move-on into independent living, so desired by Supporting People's language of transitions and transformations, comes across as much less achievable according to those practice narratives.

To some extent, Supporting People can be understood as just representing continuity of claims about the problems of institutional containment for vulnerable people and independent living movements. Those claims are consistent with, for example, Care in the Community policy under the 1990s Conservative administration in the UK (see Shaw et al., 1998). They also resonate with broader global–Western shifts in homelessness policy and practice to 'Housing First' models, which regard homeless people as resourceful human subjects whose problems are best resolved through the immediate provision of independent accommodation (alongside professional support interventions) as opposed to steps-to-resettlement approaches (Johnsen, 2010; Johnsen and Teixeira, 2010). However, there is something distinctive about both Supporting People's reparative impulses and how Windmill Project workers contested these. This is because the practice narratives covered in this chapter laid claims to professional, organisational and institutional success and failings *via* the articulation of vulnerability and claims as to what constituted care and help in modern times. The analysis developed in this chapter therefore offers a productive foundation for unpicking *processes of neoliberalisation* in marketised, networked and regulated governance regimes through consideration of how the emotions are central to practitioner actions and the social and affective relations of the state.

References

Buckingham, H. (2012) 'Capturing diversity: a typology of third sector organizations' responses to contracting based on empirical evidence from homelessness services', *Journal of Social Policy*, 41(3): 569–89.

Bury, R. (2011) 'Scale of Supporting People cuts uncovered', *Inside Housing*, 28 January 2011.

Butler, P. (2014) 'If supported housing is cut, we will see more rough sleeping and more crime', *The Guardian*, 12 February 2014.

Carmel, E. and Harlock, J. (2008) 'Instituting the "third sector" as a governable terrain: partnership, procurement and performance in the UK', *Policy and Politics*, 36(2): 155–71.

Clarke, J., Gewirtz, S. and McLaughlin, E. (2000) (eds) *New managerialism, new welfare?* Sage, London.

Clarke, J. and Newman, J. (1997) *The managerialist state.* Sage, London.

Cooper, K. (2013) 'Supporting People funding boost is a mirage for cash-strapped councils', *The Guardian*, 23 January 2013.

Department for Communities and Local Government (DCLG) (2006a) *Strong and prosperous communities: the Local Government White Paper.* HMSO, London.

Department for Communities and Local Government (DCLG) (2006b) *Homelessness code of guidance for local authorities.* HMSO, London.

Department for Communities and Local Government (DCLG) (2006c) *Places of change: tackling homelessness through the Hostels Capital Improvement Programme.* Department for Communities and Local Government, London.

Department for Communities and Local Government (DCLG) (2008) *No one left out: communities ending rough sleeping.* Department for Communities and Local Government, London.

Department for Transport, Local Government and the Regions (DTLGR) (2001) *Strong local leadership: quality public services.* HMSO, London.

Desjarlais, R. (1999) 'The makings of personhood in a shelter for people considered homeless and mentally ill', *Ethos*, 27(4): 46–89.

Dobson, R. (2009) '"Insiderness", "involvement" and emotions: impacts for methods, "knowledge" and social research', *People, Policy and Place*, 3(3): 183–95.

Dobson, R. (2011) 'Conditionality and homelessness services: "practice realities" in a drop-in centre', *Social Policy and Society*, 10(4): 547–57.

Dobson, R. and McNeill, J. (2011) 'Homelessness and housing support services: rationales and policies under New Labour', *Social Policy and Society*, 10(4): 581–89.

Harding, J. and Willett, A. (2008) 'Barriers and contradictions in the resettlement of single homeless people', *Social Policy and Society*, 7(4): 433–44.

Hoggett, P. (2001) 'Agency, rationality and social policy', *Journal of Social Policy*, 30: 37–56.

Hoggett, P. (2008) 'Relational thinking and welfare practice', in Clarke, S., Hoggett, P. and Hahn, H. (eds) *Object relations and social relations.* Karnac, London, pp. 65–85.

Hoggett, P., Mayo, M. and Miller, C. (2009) *The dilemmas of development work: ethical challenges in regeneration.* Policy Press, Bristol.

Homeless Link (2013) *Who is Supporting People now? Experiences of local authority commissioning after Supporting People.* Homeless Link, London.

House of Commons (2004) *Office of the Deputy Prime Minister, Housing, Planning, Local Government and the Regions Committee – Supporting vulnerable and older people: the Supporting People programme tenth report of session 2003–04, Volume I.* HMSO, London.

Hunt, P. (1981) 'Settling accounts with the parasite people: a critique of "A Life Apart" by E.J. Miller and G.V. Gwynne', *Disability Challenge*, Issue 1.

Hunter, S. (2003) 'A critical analysis of approaches to the concept of social identity in social policy', *Critical Social Policy*, 23(3): 322–44.

Hunter, S. (2015) *Impossible governance?* Routledge, London.

Jarrett, T. (2012) *The Supporting People programme, House of Commons research paper 12/40*. HMSO, London.

Johnsen, S. (2010) 'Residential communities for homeless people: how "inclusive", how "empowering"? A response to "Routes out of poverty and isolation for older homeless people: possible models from Poland and the UK"', *European Journal of Homelessness*, 4: 273–80.

Johnsen, S. (2014) 'Where's the "faith" in "faith-based" organizations? The evolution and practice of faith-based homelessness services in the UK', *Journal of Social Policy*, 43(2): 413–30.

Johnsen, S. and Teixeira, L. (2010) *Staircases, elevators and cycles of change: 'housing first' and other housing models for homeless people with complex support needs*. Crisis, London.

Lea, T. (2008) *Bureaucrats and bleeding hearts: indigenous health in Northern Australia*. UNSW Press, Sydney.

Lewis, J. (2005) 'New Labour's approach to the voluntary sector: independence and the meaning of partnership', *Social Policy and Society*, 4(2): 121–31.

Lipsky, M. (2010) *Street-level bureaucracy: dilemmas of the individual in public services (30th anniversary expanded edition)*. Russell Sage Foundation, New York.

McGivern, G. and Ferlie, E. (2007) 'Playing tick-box games: interrelating defences in professional appraisal', *Human Relations*, 60(9): 1361–85.

Milbourne, L. (2009) 'Remodelling the third sector: advancing collaboration or competition in community-based initiatives?', *Journal of Social Policy*, 38(2): 277–97.

Miller, C., Hoggett, P. and Mayo, M. (2006) 'The obsession with outputs: over regulation and the impact on the emotional identities of public service professionals', *International Journal of Work Organization and Emotion*, 1(4): 366–78.

Newman, J. (2000) 'Beyond the new public management? Modernizing public services', in Clarke, J., Gewirtz, S. and McLaughlin, E. (eds) *New managerialism, new welfare?* Sage, London.

Office of the Deputy Prime Minister (ODPM) (2004) *What is Supporting People?* Office of the Deputy Prime Minister, London.

Office of the Deputy Prime Minister (ODPM) (2005) *Hostels Capital Improvement Programme: policy briefing 12*. Office of the Deputy Prime Minister, London.

Office of the Deputy Prime Minister and Social Exclusion Unit (ODPM and SEU) (2004) *Tackling social exclusion: taking stock and looking to the future*. HMSO, London.

Robinson, C. (2011) *Beside one's self: homelessness felt and lived*. Syracuse University Press, New York.

Rough Sleepers Unit (2000) *Coming in from the cold: the government's strategy on rough sleeping*. Office of the Deputy Prime Minister, London.

Scanlon, C. and Adlam, J. (2008) 'Refusal, social exclusion and the cycle of rejection: a cynical analysis?', *Critical Social Policy*, 28(4): 529–49.

Shaw, I., Lambert, S. and Clapham, D. (eds) (1998) *Social care and housing, vol. 32*. Jessica Kingsley, London.

Whiteford, M. (2010) 'Hot tea, dry toast and the responsibilisation of homeless people', *Social Policy and Society*, 9(2): 193–205.

9   Fearful asymmetry

Circuits of paranoia in governing through school inspection

John Clarke

In this chapter, I explore the perverse dynamics of one field of governing relationships in England: the system of school inspection provided by OfSTED (the Office for Standards in Education). I suggest that this process, and the field of relationships through which it is conducted, are characterised by an emotional intensity at odds with conventional descriptions of rational bureaucratic organ-isation or claims about the forensic or scientific objectivity of audit and inspection processes (see Power, 1999, and Lindgren and Clarke, 2014, on the significance of 'forensic' imagery for school inspection). Yet this form of emotional intensity – what I describe as a circuit of paranoia – is also different from the forms and sites of emotion that have been of growing academic interest. The hard-nosed evaluative process of inspection differs in theory and practice from the 'therapeutic state' discerned by Nolan and others (Nolan, 1998). Nor is it a site of 'emotional labour' in which the organisation and management of the social is conducted through care or relationship work (after Hochschild, 1983). Rather, I suggest that the form of collective psychopathology visible in the school inspection regime is an unintended (though perhaps not unexpected) effect of a model of governing that seeks to promote continuous improvement, is constructed out of mistrust and surveillance and is conducted through organisational relationships that emphasise governmental, social and professional distance between the inspectors and the inspected. It is perhaps closer to Isin's understanding of the 'neurotic citizen' (2004) as a perverse consequence of neoliberal rule – the anxious subject that forms in the shadow of the incitement to be responsible, independent and empowered.

The first section of the chapter traces the creation and development of OfSTED as a system of school governing. This and the following section detail some of the ways in which the practice of school inspection in its OfSTED form has been both contentious and controversial. The following three sections explore some of the emotional dynamics associated with OfSTED as a mode of governing schooling, tracing formations of anxiety, suspicion and paranoia. I suggest that the dynamics of school inspection in England have been characterised by a form of collective psychopathology – a circuit of paranoia that operates in the whole field of relationships between individuals and organisations engaged by this process. In the conclusion, I consider what this attention to emotional dynamics – and the circuit of paranoia, in particular – adds to our understanding of governing practices and relationships.

The chapter draws on a (2010–13) comparative study of systems of school inspection in England, Scotland and Sweden led by Jenny Ozga (see the project website: www.education.oc.ac.uk/governing-by-inspection). The comparison was intended to explore different modes of governing schooling in the three different national systems and to contribute to a wider analysis of the role inspection plays in the governing of public services (Grek and Lindgren, 2014; see also a special issue of the journal *Sisyphus*).[1] The study involved interviews with around 50 actors involved in the practice of school inspection in each of the jurisdictions; in England we interviewed head teachers, education officers, school inspectors, lead inspectors and former Her Majesty Inspectors (HMIs). The study also analysed official documents (policy, guidance, inspection methodologies and school reports) as well as media reportage of school inspection. During the study it became clear that the regime centred on OfSTED was distinctive in several ways: it had been created during a particular period of state reform (the Conservative governments of the 1980s and 1990s); it had been shaped by a zealous approach to the process of public service improvement; and it was marked by a history of contention between organisations and actors within the system of governing schooling. But even allowing for this history of contention (discussed further in the following section), there seemed to be a surplus – an excess – of emotional material surrounding this regime, which made its presence felt in case-study interviews and more public forms (mass media). Such emotional excess was not visible or audible in the other two case studies: Sweden and Scotland seemed (emotionally) cooler and warmer respectively. As anthropologists have previously suggested, emotions might well have specific conditions of time, place and cultural form (e.g. Lutz, 1988; Lutz and White, 1986). However, this is not the place for a comparative study of emotional atmospheres; instead, this chapter focuses on the puzzle that the English regime represents – how to understand the surplus of emotion that seemed to swirl around this mode of governing schooling.

Contentious governing?

In 1992, OfSTED replaced the long established Her Majesty's Inspectorate of Schools (founded in 1839) as part of a larger reform of the architecture of governing in the UK undertaken by the Conservative governments of 1979–97. OfSTED was both a modernised and a modernising agency: it embodied a new approach to governing public services at a distance, and it was expected to ensure that the organisations, agents and processes that it governed became modern in their turn. The organisation was created as part of the Conservative reforms of education that promoted greater 'autonomy' for schools (or, at least, some schools), promised greater 'choice' for parents, nationalised the curriculum and aimed to overthrow approaches to teaching that were variously labelled as 'liberal', 'permissive' or 'child-centred'.

The new arrangements for governing schooling, like many other public services, were articulated around the principles of 'governing at a distance', rather than through the systems of integral government department bureaucracies (Clarke, 2014). The integral state was subjected to a programme of dispersal: multiplying the number and form of organisations involved in delivering services, and creating new organisations to direct, scrutinise and evaluate the performance of service providers (Newman and Clarke, 2009). In particular, forms of scrutiny, evaluation, audit and inspection took on increasingly important roles as means of managing dispersed or fragmented systems of provision (Power, 1999; Pollitt and Summa, 1999). Many of these organisations also adopted a view of their role as speaking for, and to, the 'consumers' of public services. In OfSTED's case, this identity was articulated in relation to the parents and pupils of schools, but also other groups of 'users' such as employers, communities, tax payers (see, for example, Clarke, 2005; Clarke et al., 2000; on OfSTED, see Clarke and Baxter, 2014).

OfSTED came into existence with the promise that every school (primary and secondary) in England would be inspected within four years, and would then receive repeated inspections. The centrality of inspection to the role and practice of OfSTED was embodied in its first corporate mission statement: 'Improvement through Inspection'. The scope of inspection also demanded a change in staffing, the core Inspectorate shrank from around 515 to 300 HMIs, with inspections to be staffed largely through subcontracted inspectors. The inspection process was contractualised and put out for tender (another common 'marketising' reform of the Thatcher governments: on OfSTED, see Lawn, 2014). This system of sub-contracting will be replaced by direct contracting of inspectors by OfSTED from September 2015.

Initially, the culture of the patrician-professional Inspectorate seemed to dominate the new organisation and its relations to government. However, the appointment of Chris Woodhead as Her Majesty's Chief Inspector (HMCI) in 1994 (he served until 2000) is viewed as changing the style of the organisation in a number of ways. Smith (2000), for example, describes him as leading a transformation of OfSTED into a 'campaigning organisation', which adopted explicit public stances on teaching methods, the quality of teachers, the curriculum and school performance. He – and the organisation – also propounded a zealous belief in the transformative power of inspection, despite limited or even contradictory evidence about its impacts (on zeal and public bureaucracies, see du Gay, 2000).

OfSTED represents a distinctive (if shifting) inspection regime, different in a number of respects from the two other national regimes that we examined (Scotland and Sweden). It differs in its institutional location in the 'machinery of government', being an 'arm's length' agency of government, separate from the Department of Education. It differs in its organisational form: the extensive contracting out of the practice of inspection (currently to three corporate providers of inspectors). It differs in the framework that shapes and informs inspection and judgement (albeit with shifting frameworks). Finally, it differs in the degree of professional and governmental distance between inspectors and the inspected. This last point is particularly significant, given that it appears to constitute school inspection in England as a peculiarly antagonistic relationship (for more on the three inspection regimes, see Grek and Lindgren, 2014).

Dogged by controversy

OfSTED has been surrounded by controversy since its creation. The controversies have moved between different issues: methodological, organisational and political. The practice of inspection – especially on OfSTED's almost industrial scale, described by Field et al. (1998: 126) as 'the bureaucratized, pressurized and subcontracted system of school inspection' – has been controversial in terms of its methodology. There are recurrent questions about the consistency of judgement between inspections and inspectors, despite attempted standardisation by handbook and training; see, for example, Penn (2002) or Sinkinson and Jones (2001) for specific examples. In our study, the head teachers interviewed recurrently posed the problem of a lack of 'consistency' in inspection practice and judgement (Baxter and Clarke, 2014). Field et al. (1998: 127) observe that: 'The process is standardized and therefore presented as objective and fair.' But, methodologically speaking, neither standardisation nor independence and impartiality guarantee reliable and comparable outcomes (Smith, 2000).

The OfSTED process of inspection has been viewed as producing perverse organisational effects. Inspection has been represented as time consuming, expensive and corrosive of trust and professional culture, and many studies have pointed to the dislocation and distraction associated with being inspected (e.g. Perryman, 2007). Several studies also refer to the performative character of the inspection process, with recurrent use (by school staff) of metaphors such as 'jumping through hoops' and 'papering over the cracks' (Plowright, 2007: 384), while Case et al. (2000: 615–17) report nominal compliance and the 'performance' of accountability and good teaching on a 'stage managed' basis. They conclude that everyone – including OfSTED – has a need to 'show you're working' (see also Clarke, 2005, and Perryman, 2009, on the performative and panoptic qualities of inspection).

The proliferation of controversy, adversarial positions, antagonistic encounters and outbursts of hostility placed OfSTED in an unusually visible and problematised position in the world of governing. Often publicly named as a schools 'watch dog', it is also sometimes condemned as an 'attack dog' (attached to the Secretary of State on a short leash, but not named in the Dangerous Dogs Act). Alternatively, it has been viewed as a 'mad dog', capable of turning nasty, or – rather differently – as a 'lap dog', excessively comfortable in its proximity to the Secretary of State for Education (the dog metaphors are borrowed from Hackett, 2001). Inspection in the OfSTED mode appears to have been considerably more controversial and antagonistic than in our other examples. What is it about the English system that appears to put more 'distance' between inspectors and schools? In the following sections, I explore the collective psychodynamics of school inspection in the OfSTED mode, suggesting that the field of relationships of inspection might be described as a form of collective psychopathology: a circuit of paranoia.

Becoming anxious: the world of inspection

The OfSTED inspection process has frequently been discussed in terms of the levels of stress and anxiety that it can generate. For example, a representative of the National Union of Teachers (NUT) said the union had perceived increasing stress among teachers, resulting from the current model of inspection:

> The current model is about getting teachers to show how they've met their targets – if they haven't done so immediately there's a very quick procedure, not to support teachers, but towards disciplinary action and dismissal. That creates a context in which teachers feel under pressure.
>
> (quoted in Ratcliffe, 2012)

In some senses, this is not surprising, since part of the ethos of OfSTED is that pressure is needed to drive up standards. Indeed, the current Chief Inspector, Sir Michael Wilshaw, has been scathing about schools (and teachers) who are 'coasting'. His concern about such schools drove one critical change in the inspection framework in 2013, which saw the replacement of the category of 'Satisfactory' with 'In need of improvement'. In a series of comments, he challenged teachers and head teachers who complained about their jobs:

> Sir Michael Wilshaw, the head of OfSTED, said that being a head teacher was a brilliant, well paid job and that school leaders had no grounds to complain . . . The comments risk further infuriating the teaching profession which has recently been told by Sir Michael that there is no stress in teaching and that staff who are out the school gates at 3.30pm should be paid less.
>
> 'I have no time for head teachers who go around moaning,' Sir Michael told heads, teachers and academics at the Institute of Education, in London. 'They have to get on and do it.'
>
> (Henry, 2012)

Commenting on such bracing interventions, the *New Stateswoman* (2012) argued that: 'The reign of Sir Michael thus far has been peppered with controversy – this is a man who likes to make strong statements and to watch the reaction.' OfSTED has proved adept at maximising media coverage of its judgements and views, with the Chief Inspector always newsworthy. However, as we shall see, such a mediatised presence for OfSTED and the Chief Inspector has proved to be a double-edged sword. For the moment, though, these exchanges and anxieties fall within the realm of contentious governing relationships – and only the question of stress hints at the question of emotion. But, as Jane Perryman (2007: 182) has argued in her study of the emotions of inspection, 'it is important when analyzing my own data to move beyond glib references to stress and look instead at the emotions within the statements'. She states that, rather than the very visible issue of stress and overwork associated with OfSTED inspections, the critical theme 'appears to be fear. The teachers are not expressing their dislike at overwork, nor complaining about stress, but there seems to be a genuine fear ... Fear of the consequences of a poor OfSTED report drives people on in terms of massive overwork, and it is the emotion of fear, not stress of overwork, that is the important reaction' (ibid.: 180).

Perryman traces a series of emotional responses among the teachers in the specific school she was studying and identifies a tendency towards 'disaffection' as the result of OfSTED inspections, condensing resentment, suspicion and a sense of being undermined as a professional (and as a person). This is an enormously suggestive study but is confined to studying the recipients of inspection. In what follows I want to suggest that these emotional states are in play across the field of relationships, even if they are differentially distributed. For example, the sense of fear and anxiety has been articulated by both an OfSTED inspector and a head teacher:

> I know you're nervous but so am I. Your nervousness is well founded. You know my judgments are going to affect your future – and might put you out of work if things go badly.
>
> (Anonymous, 2013)

> Usually the anticipation of an event is worse than the event itself. This is not my experience of OfSTED inspections. This is my fifth full inspection as a head and still my anxiety levels are high each time. I didn't sleep that night.
>   The hardest part of any inspection in some ways is the days that follow, with all the staff feeling shellshocked and exhausted. I could hardly string a sentence together. We are left dazed and battered in their wake.
>
> (Bergistra, 2012)

By comparison with the Scottish and Swedish inspection regimes, the OfSTED approach seems to place a dynamic of mistrust, suspicion and anxiety at the very heart of the process. It inflects the relationships between inspectors and inspected before, during and after the inspection.

Suspicious minds

The antagonistic relationships established in the first decade of OfSTED's work have been articulated in a culture of mutual suspicion and mistrust. The inspectorate apparently mistrusts at least some schools, teachers and local education authorities. In many respects, the proclaimed necessity of inspection (and other forms of evaluation and audit) rests on a fundamental principle of mistrust, as Onora O'Neill argued in her 2002 Reith Lectures (O'Neill, 2002). In the rise of neoclassical economic perspectives, public services have become particularly mistrusted because they are seen as relatively immune to the corrective disciplines of market forces. OfSTED, however, has become distinctively suspicious – of teachers, schools, school leaders and local authorities – being concerned that many have been evading their responsibilities. This suspicion has been reflected in continuing debates about short-notice or no-notice inspections to ensure that schools cannot conceal their true character by preparing for inspection. For example, Sir Michael Wilshaw commented that:

> OfSTED has been moving towards a position of unannounced school inspection over a period of years. I believe the time is now right for us to take that final step and make sure that for every school we visit inspectors are seeing schools as they really are in the corridors, classrooms and staffroom.
>
> (in Vasagar, 2012)

This condition of suspicion appears endemic to inspection processes that are conceived as adversarial. In return, LEAs, teachers and schools mistrust OfSTED, the Chief Inspector and the inspection teams, not least because they recognise the distrust that is at stake in the inspection process. One OfSTED lead inspector reflected carefully on the distribution of trust and mistrust in the inspection process (at the point when the 'satisfactory' judgement was being replaced by 'in need of improvement'):

> We've already moved, haven't we? We've already moved to say that outstanding schools don't need inspecting, no one's had a moan about that, have they?
>
> . . . so on the plus side they are saying 'we trust you', if you want to . . . we trust you. I think that's great. I do think if senior leaders and heads change that, well we should be risk assessing. But then at the other end of the scale, we are saying we don't trust you, satisfactory, we don't trust you, at the bottom line they are saying we don't trust you and satisfactory's not good enough.
>
> (Participant 4, lead inspector)

However, mistrust is not just a condition where there is a lack of trust: it is an active emotionally ordered relationship, involving doubt, scepticism and – where power and its consequences are at stake – fear and anxiety. In classic psychoanalytic terms, debates around schools and inspection often involve 'splitting' (the binary and absolutist distinction of Good and Bad) and projection (the phantastic imagery of bad and good people). Froggett (2002: 37) describes the psychoanalytic basis of splitting as follows:

> The splitting [of good and bad] protects the fragile developing ego by keeping the phantasy of the good apart from, and uncontaminated by, the bad. Although processes of psychic differentiation and integration will eventually allow the developing child to develop the capacity for ambivalence and the ability to relate to whole objects, splitting remains an integral part of the defensive repertoire – always the first to be mobilised when under threat. Welfare agencies, hospitals and schools are very familiar with splitting in clients who rage against a particular worker while idealising another. This

allows identification with an individual; who is protected from negative projections and becomes the bearer of hope for change; however it also defends against the need to come to terms with an imperfect and contradictory reality.

The rhetorical landscape of schooling and inspection is littered with such distinctions (about pupils, about schools and even about inspectors). Such distinctions do indeed speak to the problem of aligning classificatory/judgement systems with 'an imperfect and contradictory reality'. But the relational responses that come to operate are characteristically anxious, defensive and paranoid. As the earlier discussion of Perryman's work indicates, these responses have been most visible among the inspected, but in our study we found that – at moments – they are also true for inspectors. So one lead inspector pointed to the elaborate quality control arrangements that scrutinise the reports produced by inspectors, involving checks internal to the inspection provider organisations and then at OfSTED:

> [B]ut the real difficulty for the providers is that, the report gets read by one of the quality readers, then it comes back to me, then it goes to the school for a factual accuracy check, and it gets put together and gets sent to OfSTED for sign off.
>
> Now OfSTED say no we are not signing it off, then it becomes a key performance indicator failure for the provider, so they are paranoid about this because they get slapped, I know because I was Director of Inspections for one of the contractors in the past, you get contract action notices that will say that unless you improve this will happen and you have to get it right, and of course once you do it that way and then HMI say no it has to be like this you end up, with a formula, I think that's part of an issue, that is why schools are saying er . . . all these defences are going up.
>
> (Participant 1, lead inspector)

In this brief extract, we can see the circulation of paranoia: among inspectors (writing reports), among inspection provider organisations (submitting reports) and among schools trying to protect themselves against the application of an inspection formula. This mode of inspection creates anxiety and uncertainty. While most discussion of this focuses on the schools and teachers, I think it is worth considering how the field of antagonistic relationships also affects the inspectors (individually and in teams) and OfSTED itself. The field of relationships (OfSTED–inspection provider companies–contracted inspectors–schools) is structured by a principle of uncertainty (about judgements to be made) that induces anxiety. There are institutionalised forms of mistrust that generate defensive reactions and fear. Indeed, OfSTED has often been aggressively defensive about its approach and its judgements, and at times appears offended by criticisms. For example, in 2014 it issued a 'clarification for schools' about the inspection process, which claimed to confirm 'facts about the requirements of OfSTED and dispel[s] myths that can result in unnecessary workloads in schools' (OfSTED, 2014). In the text, the phrase

'OfSTED *does not*' (emphasis in original) recurs 16 times in two pages to disavow the requirements erroneously attributed to the OfSTED process.

Widening the circuit of paranoia?

These paranoid dynamics extend beyond the OfSTED–contracted inspection–school circuit into the wider realm of public and political action. One small, but intriguing, example was provided by the current Chief Inspector in his address to the National Governors Association in 2012:

> In my short time in the post I seem to have picked up something of a reputation and I am actually quite a nice man.
> I promise you that not everything that's been written about me is true!
>
> (Wilshaw, 2012)

This is, at least, an interesting rhetorical device – recognising a publicly contentious persona and attempting to disarm a potentially sceptical audience. But it might also be read as a paranoid defence against a hostile and judgemental world, which (he feels) is not sufficiently attentive to this 'contradictory reality'. OfSTED's defensiveness may seem strange for an apparently powerful and authoritative institution, but such features are not necessarily any defence against perceived hostility, antagonism and cruel (mis)judgement. Indeed, during 2014 the Chief Inspector and OfSTED found themselves the subjects of critical evaluations from 'sources close to government'. Two think-tanks with strong links to the Conservative Party delivered critical reports on OfSTED. *Civitas* argued that the OfSTED approach was stifling innovation in school organisation and teaching, not least because of a 'progressive' or child-centred orthodoxy at OfSTED (Peal, 2014). The Policy Exchange report claimed that the inspection process was poorly staffed and produced unreliable and inconsistent judgements, while placing undue pressure on schools (Waldegrave and Simons, 2014). In January 2014 the BBC reported on an angry Chief Inspector who feared that sources inside the Department for Education, headed by Minister of Education Michael Gove, had been briefing against him:

> Sir Michael told the *Sunday Times* he suspected the think tanks were being 'informed by the Department for Education' – 'possibly' Mr Gove's special advisers – and that he was 'displeased, shocked and outraged'.
> 'I am spitting blood over this and I want it to stop,' he said.
> Asked whether he wanted Mr Gove to call off the 'attack dogs', the newspaper reported, he replied: 'Absolutely.'
> He added: 'It does nothing for [Michael Gove's] drive or our drive to raise standards in schools.
> 'I was never intimidated as a head teacher and I do not intend to be intimidated as a chief inspector.'
>
> (BBC News, 2014)

These are classically paranoid reactions to perceived threats and slights. This is not to suggest that the threats and slights were not real or intended to cause political damage. However, the response projects fears onto shadowy but powerful others (briefers in the Department of Education) and escalates the aggressive tone ('spitting blood', 'I do not intend to be intimidated'). The anger and the projection of powerful enemies continued during 2014, with Wilshaw accusing 'vested interests' of trying to block his reforms in October (Hurst, 2014). At the same time, OfSTED was embroiled in a rather different field of political paranoia, with an investigation ('Trojan Horse') into claims that Muslim governors were taking over schools in Birmingham and elsewhere with the aim of promoting a curriculum and school ethos that deviated from 'British values'. Fears that the Birmingham experience was just the 'tip of the iceberg' in a plot to 'Islamise' British schools were voiced by lead investigator Peter Clarke (Gilligan, 2014; see also Baxter, 2014 and forthcoming).

This issue provides a further indication of the ways in which OfSTED operates within an expanding circuit of collective paranoia in which suspicion, mistrust, fear and anger are recurring features. This is not quite paranoia in the classical Freudian sense, since it is not focused on a single ego, but it catches echoes of some of the symptoms attributed to paranoia as a defence mechanism: the sense of persecution, the projection of the power of the persecutors and the reviling of the persecutor (see, for example, Freud, 1914). There are also interestingly suggestive echoes of Freud's work on the relationship between surveillance/judgement, internalised self-monitoring and perceived failure of the self to match up to the 'ideal ego' (it sounds like a model for an inspection system). However, my purpose is not to elaborate a Freudian analysis of OfSTED's mode of inspection and the social relationships in which it is enacted. Rather, I want to borrow the *idea* of paranoia (and its psychopathological echoes) as a way of describing this field of relationships and the circuits through which they are connected. This draws more on the socialisation of paranoia as a concept for studying organisational or group dynamics (see, for example, Marcus, 1994, and Mirowsky, 1985). Again, though, I think there is a difference between dealing with a field of relationships rather than one subject, even a collective one (an organisation or group). Here, it is the field of relationships between agents and agencies that creates a circuit of paranoia as the dynamic of judgement and anxiety (inflected by power and its effects) enrolls all of the different actors. Rather than paranoia as a state of mind, or an emotional state, it is lodged in the relationships and practices of this field of relationships as a regime of governing.

Conclusion: emotional states and states of emotion?

This is a preliminary speculation and I am not sure where it might lead. Emotional intensity – and the circuit of paranoia that seems to suffuse the field of relationships – certainly mark the English inspection regime as different from those in Scotland and Sweden. Both of these certainly contain elements of anxiety and moments of suspicion, but they do not seem to dominate the field of relationships in the same way – and are tempered by other modes of interacting. So how might the intensity of the emotional register associated with OfSTED inspection be explained?

One starting point is to recognise that this inspection regime has produced a field of antagonistic relationships, which are reflected in the intensity of feeling in which positions, experiences and aspirations are expressed. But 'reflection' is a poor conceptual tool for capturing these dynamics. Alternatively, this excess of emotion might be explained in terms of different rhetorical possibilities: what ways of speaking are available to the different actors that constitute this 'system': inspectors, teachers, head teachers, local authorities, parents, etc. (and the children/ pupils who are typically spoken for)? What cultural resources are available to represent personal, professional and organisational experiences in publicly acceptable, compelling or persuasive ways? Such resources might include affective vocabularies (stress, anxiety, fear, despair, anger and so on). There is a growing interest in the role and significance of discursive vocabularies of justification (Boltanski and Thévenot, 2006; see also the earlier work of Scott and Lyman, 1968, on accounts). Such perspectives take a performative view of language – stressing the ways in which specific vocabularies legitimate, enable and bring into being particular courses of action. However, these approaches tend to emphasise a rational view of vocabularies, rather than considering what affective/emotional repertoires may legitimate or enable. In this case, I suggest, it may be possible to develop an analysis of how the English inspection regime has accumulated a vocabulary of emotion through which a series of antagonistic relationships may be legitimately represented – and called into question. So, talk of stress, anxiety, fear, suspicion and mistrust draws on a collective vocabulary through which (some) actors located in this field may articulate their experiences and challenge the dominant instrumentalising vocabulary of inspection and improvement. There is, I think, some potential in thinking through the ways in which this emotional repertoire is mobilised, legitimised and deployed by different actors within the field. And yet, I am left with a question about what would happen if we took this richness of *feeling* about the relationships and practices of inspection more seriously?

An alternative starting point is the emerging literature on the psychosocial dynamics of public services, organisations and work (for example, Froggett, 2002; Hoggett, 2000; Long, 2008). They point to the signs of stress, strain and perverse dynamics in organisational and occupational settings. Long, indeed, has written of *perverse* organisations, though, sadly, not in terms of paranoia as a common organisational dynamic. Here I want to stretch this interest in collective or social psychodynamics beyond the specific organisation or occupation towards thinking of school inspection as a field of relationships. In the English variant, as constituted through OfSTED's central role, this field contains – and perhaps is animated by – a shared paranoid sensibility. This is ingrained in habits and repetitions that have become normalised in circuits of recursive practices, such that anxiety, fear, suspicion and mistrust are recurrent, if not dominant, dispositions of the field.

It matters, I think, to be careful about the specificities of this argument. It is not my view that all inspection systems are characterised by paranoia, or that all government/governance arrangements produce collective psychopathologies. Rather I want to suggest that this particular field of relationships has been constituted in such a way as to incline agents within the circuit to experience paranoid reactions. Wetherell's insistence on thinking about 'affective practices' is, I think, helpful in this respect. She argues (2013: 235–6) that:

> An affective practice approach, then, takes as its focus and its units of analysis patterns and cycles of activity that at a particular historical moment have become 'emotionalised' (understood through the conventional categories and vocabularies of emotion) . . .
> An affective practice typically pulls together or orders in relation to each other patterns of body/brain activity, patterns of meaning-making, feelings, perceptions, cognition and memories, interactional potentialities and routines, forms of accountability, appraisals and evaluations, subject positions and histories of relationships.

This is a suggestive way of thinking about the OfSTED inspection regime as a circuit of paranoia, because it embeds the study of affect/emotion in specific configurations of relationships and practices. Her call for attention to 'cycles of activity' that have become 'emotionalised' addresses the distinctive qualities of this particular field of relationships – and the practices in which it is embodied. The actual and anticipated practice of inspection (and its reciprocal – being inspected) is enacted in patterns of activity and patterns of meaning-making that are simultaneously routinised and highly charged because of the political, professional, organisational and personal stakes. In the case of the OfSTED circuit, it is also clear that the unequal relations of power – the 'fearful asymmetry' of the chapter title – experienced by different agents and agencies within this field are crucial elements in determining how this paranoid field of relationships has been produced, reproduced and inhabited.

Notes

1 Volume 2, issue 1: http://revistas.rcaap.pt/sisyphus/issue/view/300 (accessed 11 May 2015).

References

All website URLs were accessed on 11 May 2015.

Anonymous (2013) 'What I'm really thinking: the OfSTED Inspector', *The Guardian*, 2 February, at: www.guardian.co.uk/lifeandstyle/2013/feb/02/what-really-thinking-ofsted-inspector.

Baxter, J. (2014) 'Trojan Horse: snap school inspections will not solve wider governance issues', *The Conversation*, 10 June 2014, at: http://theconversation.com/trojan-horse-snap-school-inspections-will-not-solve-wider-governance-issues-27824.

Baxter, J. (forthcoming) *Governing schools: policy, politics and practices post Trojan Horse Affair.* Policy Press, Bristol.

Baxter, J. and Clarke, J. (2014) 'Knowledge, authority and judgement: the changing practices of school inspection in England', *Sisyphus: Journal of Education*, 2(1): 106–27.

BBC News (2014) 'Sir Michael Wilshaw "spitting blood" over attack', 26 January 2014, at: www.bbc.co.uk/news/education-25900547.

Bergistra (2012) 'OfSTED is the last thing you need when a pupil is having a tantrum', *The Guardian*, 17 December 2012, at: www.guardian.co.uk/education/2012/dec/17/headteacher-on-a-knife-edge.

Boltanski, L. and Thévenot, L. (2006) *On justification: economies of worth* (translated by C. Porter). Princeton University Press, Princeton, NJ.

Case, P., Case, S. and Catling, S. (2000) 'Please show you're working: a critical assessment of the impact of OSTED inspection on primary teachers', *British Journal of Sociology of Education*, 21(4): 605–21.

Clarke, J. (2005) 'Performing for the public? Desire, doubt and the governance of public services', in du Gay, P. (ed.) *The value of bureaucracy*. Oxford University Press, Oxford, pp. 211–32.

Clarke, J. (2014) 'Inspection: governing at a distance', in Grek, S. and Lindgren, J. (eds) *Governing by inspection*. Routledge, London.

Clarke, J. and Baxter, J. (2014) 'Satisfactory progress? Keywords in English school inspection', *Education Inquiry*, 5(4): 481–96.

Clarke, J., Gewirtz, S., Hughes, G. and Humphrey, J. (2000) 'Guarding the public interest? Auditing public services', in Clarke, J., Gewirtz, S. and McLaughlin, E. (eds), *New managerialism, new welfare?* Sage, London.

du Gay, P. (2000) *In praise of bureaucracy: Weber–organization–ethics*. Sage, London.

Field, C., Greenstreet, D., Kusel, P. and Parsons, C. (1998) 'OFSTED inspection reports and the language of educational improvement', *Evaluation and Research in Education*, 12(3): 125–39.

Freud, S. (1914) 'On Narcissism', in Strachey, J. (ed.) *The standard edition of the complete psychological works of Sigmund Freud, volume XIV (1914–1916): On the history of the psycho-analytic movement, papers on metapsychology and other works*, pp. 67–102.

Froggett, L. (2002) *Love, hate and welfare: psychosocial approaches to policy and practice.* Policy Press, Bristol.

Gilligan, A. (2014) 'Trojan Horse "just the tip of the iceberg"', *Daily Telegraph*, 10 October 2014, at: www.telegraph.co.uk/education/educationnews/11157116/Trojan-Horse-just-the-tip-of-the-iceberg.html.

Grek, S. and Lindgren J. (eds) (2014) *Governing by inspection*. Routledge, London.

Hackett, R. (2001) 'News media and civic equality: watchdogs, mad dogs or lap dogs?', in Broadbent, E. (ed.) *Democratic equality: what went wrong*? University of Toronto Press, Toronto, pp. 197–213.

Henry, J. (2012) 'OfSTED head Sir Michael Wilshaw tells heads to "stop moaning"', *Daily Telegraph*, 26 December 2012, at: www.telegraph.co.uk/education/educationnews/9685479/OfSTED-head-Sir-Michael-Wilshaw-tells-head-teachers-to-stop-moaning.html.

Hochschild, A. (1983) *The managed heart*. University of California Press, San Fransisco.

Hoggett, P. (2000) *Emotional life and the politics of welfare*. Macmillan, Basingstoke.

Hurst, G. (2014) 'Wilshaw accuses critics of running a smear campaign', *The Times*, 11 October 2014, at: www.thetimes.co.uk/tto/education/article4233689.ece.

Isin, E. (2004) 'The neurotic citizen', *Citizenship Studies*, 8(3): 217–35.

Lawn, M. (2014) 'Outsourcing the governing of education: the contemporary inspection of schooling in England', *Sisyphus Journal of Education*, 2(1): 88–105.

Lindgren, J. and Clarke, J. (2014) 'The (C)SI effect – school inspection as crime scene investigation', in Lawn, M. and Normand, R. (eds) *Shaping European education: interdisciplinary approaches*. Routledge, London.

Long, S. (2008) *The perverse organization and its deadly sins*. Karnac Books, London.

Lutz, C. (1988) *Unnatural emotions: everyday sentiments on a Micronesian atoll and their challenge to Western theory*. University of Chicago Press, Chicago, IL.

Lutz, C. and White, G.M. (1986) 'The anthropology of emotions', *Annual Review of Anthropology*, 15: 405–36.

Marcus, E.R. (1994) 'Paranoid symbol formation in social organizations', in Oldham, J.M. and Bone, S. (eds) *Paranoia: new psychoanalytic perspectives*. International Universities Press, Madison, CT, pp. 81–94.

Mirowsky, J. (1985) 'Disorder and its context: paranoid beliefs as thematic elements of thought problems, hallucinations, and delusions under threatening social conditions', in Greenley, J.R. (ed.) *Research in community mental health*. JAI Press, Greenwich, CT, pp. 185–204.

New Stateswoman (2012) 'Sir Michael Wilshaw and the 3pm myth', at: http://thenewstateswoman.wordpress.com/2012/09/22/sir-michael-wilshaw-and-the-3pm-myth/.

Newman, J. and Clarke, J. (2009) *Publics, politics and power: remaking the public in public services*. Sage, London.

Nolan, J. (1998) *The therapeutic state: justifying government at century's end*. New York University Press, New York.

OfSTED (2014) *OfSTED inspections: clarification for schools*. OfSTED, London, at: www.gov.uk/government/publications/ofsted-inspections-clarification-for-schools.

O'Neill, O. (2002) *A question of trust: the BBC Reith lectures 2002*. Cambridge University Press, Cambridge.

Peal, R. (2014) *Playing the game: the enduring influence of the preferred OfSTED teaching style*. Civitas, London, at: www.civitas.org.uk/pdf/PlayingtheGame.pdf.

Penn, H. (2002) '"Maintains a good pace to lessons": inconsistencies and contextual factors affecting OSTED inspections of nursery schools', *British Educational Research Journal*, 28(6): 879–88.

Perryman, J. (2007) 'Inspection and emotion', *Cambridge Journal of Education*, 37(2): 173–90.

Perryman, J. (2009) 'Inspection and the fabrication of professional and performative process', *Journal of Education Policy*, 24(5): 611–31.

Plowright, D. (2007) 'Self-evaluation and OfSTED inspection: developing an integrative model of school improvement', *Educational Management Administration and Leadership*, 35(3): 373–93.

Pollitt, C. and Summa, H. (1999) 'Performance audit and public management reform', in Pollitt, C., Girre, X., Lonsdale, J., Mul, R., Summa, H. and Waerness, M. (eds) *Performance or compliance? Performance audit and public management in five countries*. Oxford University Press, Oxford, pp. 1–14.

Power, M. (1999) *The audit society: rituals of verification*. Oxford University Press, Oxford.

Ratcliffe, R. (2012) 'Rise in teachers off work with stress – and unions warn of worse to come', *The Guardian*, 26 December 2012, at: www.guardian.co.uk/education/2012/dec/26/teachers-stress-unions-strike?INTCMP=SRCH.

Scott, M. and Lyman, S. (1968) 'Accounts', *American Sociological Review*, 33(1): 46–62.

Sinkinson, A. and Jones, K. (2001) 'The validity and reliability of OFSTED judgements of the quality of secondary mathematics initial teacher education courses', *Cambridge Journal of Education*, 31(2): 221–37.

Smith, G. (2000) 'Research and inspection: HMI and OFSTED, 1981–1996: a commentary', *Oxford Review of Education*, 26(3): 333–52.

Vasagar, J. (2012) 'Schools face no-notice OfSTED inspections', *The Guardian*, 10 January 2012, at: www.guardian.co.uk/education/2012/jan/10/schools-no-notice-ofsted-inspections.

Waldegrave, H. and Simons, J. (2014) *Watching the watchmen*. Policy Exchange, London, at: www.policyexchange.org.uk/images/publications/watching%20the%20watchmen.pdf.

Wetherell, M. (2013) 'Feeling rules, atmospheres and affective practices: some reflections on the analysis of emotional episodes', in Maxwell, C. and Aggleton, P. (eds) *Privilege, agency and affect: understanding the production and effects of action*. Palgrave Macmillan, Basingstoke, pp. 221–39.

Wilshaw, M. (2012) *Strong governance: learning from the best. Speech to National Governors Association Conference*, 16 June. OfSTED, London.

10 Troubling feelings in family policy and interventions

Eleanor Jupp

Introduction

In this chapter I will consider the emotions surrounding forms of UK government family support, especially support aimed at poor or disadvantaged families. Such policy programmes and interventions involve 'emotional governance' at a number of levels, including explicit attempts to shape the behaviour of family members (see Lister, 2003) in what might be understood as a means of fashioning the intimate subjectivities of citizens (Rose, 1989). The focus here though is mostly slightly different, in that I am interested in the broader affective atmospheres of policy and practice which shape such interventions, especially the feelings of professionals about the families with whom they are working. In this chapter I suggest that a particular emotional discourse or affective 'pattern' (Wetherell, 2012) can be discerned across policy and popular culture about poor families within contemporary austerity Britain, which I argue has an increasingly stigmatising, even hysterical tone.

However, following my previous analysis of the emotional geographies of community work (Jupp, 2013a), I argue that, as well as producing critical appraisals of these affective *patterns* of policy and popular discourse, research should also pay attention to the lived encounters between professionals and poor families, and the particularity of affective *practices* (Wetherell, 2012) involved. As Wetherell (2012: 119) insists, emotions are embedded in complex social contexts, or geographies, 'the messiness of social life' where wider affective patterns not only circulate but are fragmented and reworked through practices.

My interest in emotions in relation to governance and state intervention is therefore not only in the emotions evoked by policy and popular culture, but also the emotions involved in the relational, complex and ambivalent nature of inter-vention (Jupp, 2013a), especially with disadvantaged families and communities. Such ambivalence stems in part from the need for practitioners to navigate a shifting (emotional) context of state policy and programmes and reconcile this with the lived experiences and (emotional) needs of the families with whom they work, resulting in 'uncomfortable' positions (Jones, 2013) and evaluations of families. My broader theoretical purpose therefore is in developing an account of the geographies of emotional governance across both the wider geographies of policy and media *and* the intimate spaces and places of service delivery and encounter. In straddling these two spheres I seek to understand the political significance of emotions, seeing them as both critical object of scrutiny but also potentially generative of new political and ethical commitments. This position therefore goes against Pile's (2010) observations of a cleavage within human geography between critical research on the politics of affect and feminist approaches to care and emotions in practice. In order to pursue this project I suggest that researchers need to get in amongst and up close to the spaces and places in which policy is implemented, to understand the textures of emotions as embodied in particular subjects and contexts.

In this chapter therefore I move between discussing 'emotions', 'feelings' and 'affective atmospheres', all of which I argue point to embodied aspects of human experience and interaction. The chapter points to the ways in which emotions and feelings are caught up with judgements, evaluations and ethics. I also draw on a range of methods and theoretical lenses, in order to invite reflection on the methodological and theoretical resources needed to research emotions and governance. Following a vignette drawing on my experiences, the chapter begins with a critical reading of family policy during the previous UK Labour government and under current conditions of 'austerity' (see Jupp, 2016). In undertaking this reading, I pay attention to expressed and evoked emotions within policy and political discourse, and also in popular culture and media representations. I use the idea of a particular emotional discourse or affective patterning (Wetherell, 2012) around poor families in current UK social policy. In order to explore how such policy actually 'feels' when it is implemented, I also discuss my research in four UK Children's Centres (Jupp, 2012, 2013b), which looked at how staff interact with users and negotiate shifts in policy programmes. The research involved interviews with staff and users as well as ethnographic observations during 'drop-in play' sessions at the centres. Towards the end of the chapter I take a more theoretically informed cut through this field, using feminist care ethics to develop ideas about the vulnerable self and a more compassionate approach to understanding poor families, linked to a detailed empirical and 'concrete' attention to affective practices and the geographies of care.

In exploring this empirical attentiveness, I propose that researchers themselves need to be attuned to the affective geographies of different settings, and their own implications in them (England, 1994). As a way in to this discussion, and as a gesture in the direction of this approach, in the following section I will begin with a personal account of feelings and affect at a visit to see a health professional with my own young son.

Encountering emotions at the health visitors' clinic

When my son was little, I took him to a health centre to ask a health visitor[1] for advice about his eating: he was one and would only eat toast. I must have been desperate as this isn't the kind of advice I seek out readily. My memory of the occasion is hazy (as memories connected with young children often are). We saw the health visitor in a large and impersonal room where other families were talking to other health visitors. Some further context: we had just moved to a new city for my work, I was struggling to combine the new job and being a mother, and both my son and I hated the expensive but depressing nursery in which he spent his days. My son was being rather 'lively' and darting around the room. The health visitor seemed very taken with him though. She did not seem at all concerned about his eating and suggested giving him sausages, although she admitted that they were not very healthy. While our conversation was going on, another family was talking to another health visitor across the room. The little girl was around two years old and dressed in pyjamas; she was also wandering round the room while her parents spoke to the health visitor. I could hear her parents telling the health visitor that their daughter refused to get dressed in the morning. From their clothes and demeanour I could tell that the family was not well off. My health visitor leaned in to speak to me in a whisper. 'Look at that,' she said, 'isn't it a shame? Not a happy situation there, not happy at all. Not like this lovely one' (gesturing to my son).

I don't think I responded at all, to my own shame because this made me complicit in the health visitor's judgement of the family, which was pity but also something stronger and more disturbing, more like disgust. She obviously felt it strongly enough to feel compelled to share it with me, a stranger but someone whom she assumed would share her feelings – because of – class? The way I spoke? The way my own child behaved? It was clear that her feelings were tied up with a recognition and evaluation of both my and the other family's capacities as parents. On the one hand, I had been recognised as a competent and resourceful parent. But, on the other hand (because of this?), I hadn't been given the advice or support I needed. In recognising the other family as needy and in need of support, she had made a wider judgement about their overall capacities and competences. Maybe her intention was that I would feel reassured by the comparison. I wondered what had been said to the other family. Did such powerful feelings and judgements inevitably shape their interactions with the health visitor, or could the specificities of their situation also reshape these judgements?

Since this incident all those years ago, as my children have grown up I have had more odd and often awkward encounters with health visitors, doctors, playworkers, nurses, speech therapists, teachers and more. Sometimes I have found myself being judged and found wanting as a parent, and those memories sting. Of more concern for this chapter though are the ways in which support is given to and received by disadvantaged parents and families, and the affective atmosphere which might shape these encounters. I begin with an account of emotions and judgements as part of policy and political and media discourses and how these might produce feelings of professionals, before considering some methodological, theoretical and ethical frameworks that could allow for the emergence of a different type of emotionally imbued politics around disadvantaged families.

The shifting emotional landscape of UK family policy:
New Labour and beyond

Although interactions involving young children might be understood as instinctively emotive and emotional, family policy interventions in the UK are actually increasingly 'measured' by ostensibly scientific objective, rather than qualitative, indicators (Rutter, 2014), for example through the use of 'randomised controlled trials' linked to paradigms of medical research (see Featherstone et al., 2014: 62–5). Arguably shaped by the demands of a shrinking state to 'invest' only in forms of support that unequivocally 'work', such a discourse of interventions is linked to American and other international programmes or toolkits which can be delivered supposedly across different contexts in a standardised way (see Lea et al., chapter 7 in this volume). Examples of relevance to this chapter would be parenting training programmes, for example 'Triple P' (an Australian programme) and 'Incredible Years' (developed in the US). These programmes are 'evidence-based' and tested through scientific trials, offering up toolkits of methods on a commercial basis.

Viewed from such a perspective, delivering family support and interventions therefore seems a dispassionate, almost technical exercise. However, as other chapters in this volume attest (Chapter 9 by Clarke and Chapter 2 by Newman), an insistence on the technical and managerial within policy can be interwoven with highly charged emotional dynamics. This is not a case of rational, evidence-based policy conflicting with the emotional demands of 'the real world', but rather that policy discourses and programmes are themselves both caught up in and productive of emotional dynamics and affective atmospheres. In what follows I seek to trace the shifting emotional patterns around UK family policy, and also consider what these shifts might mean for the feelings of professionals.

The Labour UK government (1997–2010) developed a more systematic and nationwide approach to family policy than previous governments, especially around 'early intervention' in the lives of young disadvantaged children. Lister (2003) argues that this is part of a 'social investment state' approach, whereby spending welfare money in the 'Early Years' saves on future spending. Underpinned by detailed policy frameworks such as the 'Early Years Foundation Stage', outlining competencies and skills for pre-school children, and Every Child Matters, providing an overarching rationale and framework for all interventions with children, new programmes were developed such as Sure Start, Neighbourhood Nurseries and then a more overarching programme of Children's Centres. Sure Start programmes provided multi-service support to children under five and their families, especially families in poor communities, through the development of Children's Centres on a local basis. The Sure Start programme has subsequently been criticised by its own architects for being overly managerial and target driven and not taking sufficient account of the importance of positive emotions and care in developing relationships with families (Gerhardt et al., 2011).

This lack of perceived 'warm' emotions may be partly due to the fact that the programme sought to improve outcomes for children by compelling parents or encouraging them to act in certain ways. The programme therefore also spoke to a wider set of feelings around 'parenting', across both policy and the media and in popular culture, as a set of practices and skills which could be learnt. Gambles (2010) analyses such a 'structure of feeling' across online media such as Mumsnet and TV programmes such as Channel 4's *Supernanny*, as well as government policy, during this period. As 'parenting' entered the popular and political imagination, it spoke strongly of class differences in terms of who was seen as able to provide the kind of 'parenting' that should be aspired to (Gillies, 2007; Jensen and Tyler, 2012).

Indeed, the idea of particular families as problematic because of an inability to parent effectively was present within the policy discourses not only of Sure Start but also of other aspects of New Labour government policy. These included the so-called 'Respect' Agenda and associated use of 'anti-social behaviour orders' and Acceptable Behaviour Contracts to curb the behaviour of problem families (Jupp, 2010). Such policy discourses circulated around particular figures and also neighbourhoods, where stigmatising emotions and feelings 'stuck', in Ahmed's terms (Ahmed, 2004). These included the idea of 'neighbours from hell' (Field, 2003) whose behaviour, rather than the wider social and economic context, needed to be addressed. It is worth noting that alongside these initiatives the Labour government also introduced a whole series of policies which sought to tackle issues and living conditions within deprived neighbourhoods in a more holistic way (Lupton, 2003), creating a hybrid and ambivalent policy environment. This ambivalence is discussed below via research undertaken in Sure Start Children's Centres, after a consideration of the changing emotional landscapes of family policy under conditions of austerity.

Atmospheres of austerity

In 2010, with a change to a Conservative-led coalition government in the UK, a number of factors can be seen as contributing to a shift of mood or affective atmosphere that has heightened the emotions surrounding parenting and poor families (Featherstone et al., 2013; Jensen and Tyler, 2012; Dermott, 2012). The first of these is the condition of 'austerity' in itself (Clarke and Newman, 2012), which has involved extensive cuts to welfare spending. Reform of key benefits, such as housing benefit and disability living allowance, has been accompanied by policy pronouncements about the need to be 'tougher' on poor families and to discourage such dependency, with paid work and self-sufficiency both emotionally and financially promoted as a desirable norm. Clayton et al. (2015a: 25), focusing on the emotions of 'austerity discourses', discuss the production of a new consensus about the need for welfare cuts and 'the stirring of resentment, but also guilt and shame', in relation to poor families. Furthermore, as Clayton and colleagues also argue, austerity has increasingly sought to 'privatise' responses to poverty and insecurity, by instilling 'resilience' and self-sufficiency within families and households (see also Jupp, 2016).

As Jensen (2014) and others (Allen et al., 2014) argue, these UK policy discourses have been supported and bolstered by new genres of documentary media output, which depict in lurid detail the lives of impoverished members of families and communities, including *Benefits Street* (Channel 4), *Skint* (Channel 4), *On Benefits and Proud* (Channel 5) and more. Dubbed 'poverty porn', Jensen argues that such programmes manufacture 'a new commonsense' around the need for welfare reform and cuts. This is achieved through the use of visual, visceral and affective cues around impoverished bodies and material cultures: 'the sofa abandoned in the street, piles of windswept rubbish . . . tins of cheap lager' (Jensen, 2014: 3), which is pornographic in that it works to 'provoke an emotional sensation through a repetitive and affective encounter with the television screen'. The effect of this genre of television is to produce, perhaps paradoxically, not an understanding of or empathy with the families and their situations portrayed, but a greater distance or 'othering' of these families.

Indeed, from the perspective of the circulation of emotions and affect, an interesting aspect of the debates surrounding some of these programmes has been the insistence from programme makers that they intended to produce programmes that would elicit sympathy for their subjects (Allen et al., 2014). Whether or not this is genuine, it points to the sense that the flow of emotions around cultural representations depends not only on their contents but also on the wider geographies in which those representations circulate. A context of social media, which can be understood as a space of particularly excessive emotions and forms of judgement and evaluation, can enable both feelings and judgements to be heightened and solidified, although of course there is also always the possibility of subversion in these responses (Allen et al., 2014).

Beyond the general moods and affective atmospheres of austerity (Clayton et al., 2015), a more specific focus on family policy under the current government emerged in the wake of the riots in UK cities in August 2011. Prime Minister David Cameron made an (emotive) speech about the need for a moral 'fightback' in the face of the behaviour on display, and explicitly drew links with perceived poor parenting and family breakdown:

> I don't doubt that many of the rioters out last week have no father at home. Perhaps they come from one of the neighbourhoods where it's standard for children to have a mum and not a dad . . . where it's normal for young men to grow up without a male role model, looking to the streets for their father figures, filled up with rage and anger.
>
> (Cameron, 2011)

Cameron also announced a wave of new or reinvigorated policy initiatives, including a reworked version of an existing programme aimed at non-working families that became known as the 'Troubled Families' programme. As the quote above makes clear, from the start the programme was presented in an explicitly emotive set of discourses around family breakdown and the threat to a wider moral order that this poses. The programme involves a range of ways of working which seek to 'turn around' the lives of these families. The full programme was launched on the back of a report (DCLG, 2012) based on interviews by the programme lead, Louise Casey, with a very small number of families participating in a family intervention service. Interestingly, on one level this kind of research (which flies in the face of the paradigms of evidence-based policy discussed above) might appear to open up understanding of the realities of life for disadvantaged families. Yet, in common with the 'poverty porn' media discussed above, the report frames accounts of the families' lives in a lurid way and appears to lack compassion and understanding about their broader circumstances. The report has indeed been subsequently much criticised in relation to the failure to obtain informed consent from the families involved (Bailey, 2012) in ways which echo the controversy over the depiction of the families in *Benefits Street* (Allen et al., 2014).

Indeed, whilst the report describes the individual circumstances of the families in some detail, it also insists on characterising them collectively as a group of families that share (particularly problematic) characteristics:

> The most striking common theme that families described was the history of sexual and physical abuse, often going back generations; the involvement of the care system in the lives of both parents and their children, parents having children very young, those parents being involved in violent relationships, and the children going on to have behavioural problems, leading to exclusion from school, anti-social behaviour and crime.
>
> (DCLG, 2012: 1)

The report goes on to discuss in some detail abortions, domestic violence, physical and sexual abuse and depression. Levitas (2014) therefore identifies in the policy discourses around 'Troubled Families' the evocation of a 'burgeoning dysfunctional underclass resistant to reform'. Such a group is seen not to reflect wider social and economic problems but to be the root cause of them. Interestingly, as discussed above, such emotive language within the programme sits alongside a focus on quantitative indicators, 'payment by results' and measurable outcomes in 'turning around' these families. As already noted, this kind of policy approach speaks to the (emotional) politics of austerity, in which the rationales for wider forms of welfare and support have been swept aside, leading to reliance on narrow conceptions of changing the behaviour of certain groups, blamed for economic as well as moral instability. Indeed Eric Pickles, the minister currently in charge of this programme, recently stated that:

> The Troubled Families programme demonstrates exactly what our long-term economic plan means for people . . . more economic stability for taxpayers, as we reduce the bills for social failure and get this country living within its means.
>
> (DCLG, 2014)

An emotionally heightened, individually stigmatising and even hysterical tone around problematic families has also been identified within a number of recent high-profile child abuse cases, whereby 'blame' has also been apportioned to public sector professionals, particularly social workers, seen on the one hand as overly managerial and bureaucratic and on the other as overly liberal, incompetent and colluding with abuse (Warner, 2013, 2014). Warner (2013) analyses the 'emotional politics' of the 'Baby P'[2] case involving abuse and neglect of a small child, whereby tabloid discourses in particular stigmatised not only the parents through, for example, lurid accounts of the conditions of the house where he lived, but also the social workers involved, seen as blinded both by 'political correctness' and 'tick-box culture'. Warner discusses the 'moralising' nature of emotions in the public sphere, colouring public feelings about and judgements of both an underclass of certain families and the state professionals who engage with them.

Working the spaces of family policy

Thus far in this chapter I have evoked a range of emotions involved in social policy concerned with poor families. I have wanted to show how powerful 'affective patterns' (Wetherell, 2012) are shaped by policy and media discourses. In these next sections I develop a discussion of the extent to which state professionals' work is shaped by these affective patterns, but also consider how they might be contested, especially through the lens of feminist theory.

Indeed, against the background of an increasingly unequal society in which the experiences of the poor are increasingly 'distant' from those of others (including professionals), Featherstone and colleagues (2014: 12) point to the significance of the wider processes of 'othering' poor families (such as in the policy discourses and media discussed above) in understanding the feelings and practice of social workers. Clayton et al. (2015b) also use the idea of 'distancing' to discuss how new governance arrangements under austerity, including a culture of contracting out of services and of competition between providers, have shaped increasingly fragmented relationships between services and service users. Indeed, although we may imagine that professionals are somehow removed from popular cultural currents, they are also members of society and as such reflect the views and feelings shaping wider society. Moreover, professionals are also often parents themselves. Whilst it is true, as I go on to discuss, that public sector workers' more personal experiences and identifications might lead to more 'progressive' forms of practice than those shaped by policy (Barnes and Prior, 2009; Jones, 2013), within such an emotionally charged field as parenting the opposite may also be true.

To return to the Sure Start programme initiated under the Labour government, although this was a universal programme across England, my own research[3] (see Jupp, 2012, 2013b) has shown that in practice it often involved a reshaping of existing facilities, staff practices and programmes in response to Sure Start funding on a local basis, creating a varied geography of implementation. As indicated in my discussion above, this shift was often experienced as part of a shift away from more open-ended 'family support' to a 'child-centred' approach which involved practitioners in more controlling relationships with parents in attempts to elicit better parenting. To a variable extent, centres where the programme was developed worked towards a range of outcomes and targets for children and parents, for example in relation to breastfeeding babies and 'school-readiness' (linked to communication skills) for older children. In my research a number of professionals spoke of the need or desire to become 'tougher' in their interactions with parents within the new policy landscape, within an understanding that the norms of parenting within the neighbourhoods where they were based needed to be changed. As one worker in a centre (which had been a family support centre before becoming a Sure Start centre) said to me:

> We want to give people a way out of the culture around here . . . We're all nice people, our natural inclination is to be very accommodating . . . but in order to help people we have needed to be clearer about *our* rules and *our* culture.
>
> (Kevin, Park Children's Centre worker)[4]

The reference to 'culture' I took to mean perceived cultures of parenting and family life prevalent within the deprived neighbourhood where the centre was based. The shifts in practice referred to included removing sofas so that parents could not sit down but rather would interact with children more actively, becoming stricter about the behaviour of both parents and children, and finding ways to compel parents to take part in activities such as music sessions.

As this quote and others attest, these shifts were experienced as emotionally complex. There was a desire among workers on the one hand to hold on to core values of being supportive to poor families, but on the other to shift their practices (and values) towards the goals of current policy around reshaping parenting in particular ways. The extent to which different Children's Centres and sets of professionals embraced the approaches of Sure Start was however variable, as the following quote demonstrates, whereby a centre user at one centre (the Cornerspace) and part-time worker at another centre (West Hill) reflects on such different approaches:

> At West Hill it's more about 'learning to play with your children' so that if your child falls over and hurts themselves they'll [*i.e. staff*] come and tell you but you have to go outside and get them [i.e. *the child*]. Whereas here [*i.e. Cornerspace*] the staff will go and get your child and give them a cuddle. At West Hill it's structured by government guidelines, here it's more relaxed.

Other features of the affective atmosphere at the Cornerspace included a more relaxed attitude to parental behaviour, such as allowing them to smoke, a lack of official monitoring of users of the space (for example not requiring them to sign in) and the provision of cakes and other 'unhealthy' food, which apparently would not have been allowed in other Children's Centres. One worker at the Cornerspace centre told me:

There's a collective approach to childcare here, if you need to put your feet up and read a magazine for ten minutes that's fine . . . we're also not very strict about healthy eating . . . some children here just need calories and some of our mums, they need cakes!

This 'collective approach' to childcare which allows parents a rest can be understood as in direct opposition to the approach at the Park whereby sofas were removed specifically to prevent this happening. Indeed this quote is indicative of a different ethical approach expressed by staff to interacting with families in evidence at the Cornerspace, based on an ethic of care and a more pragmatic evaluation of individual families' needs and feelings. Elsewhere (Jupp, 2013b) I discuss in more detail differential emotional geographies of Children's Centres, and how these produced different affective practices. Through my brief discussion of the research here, I hope to have indicated some of the ways in which the wider 'affective patterns' of policy are both played out but also potentially reworked within the 'affective practices' of spaces of policy implementation.

Indeed. as this and other research has shown, professionals often draw on longer-term commitments and identifications in opposition to policy demands (Jupp, 2013a), for example commitments to certain place-based communities which may indeed increase within a context of austerity. Clayton et al.'s (2015b) research with public sector workers in the North East found care and commit-ment to service users increasing in the face of ongoing poverty and cuts, despite huge pressures and stress experienced by the workers as their services came under enormous financial pressure and shifting governance arrangements. Media reports (see Tickle, 2014) have found evidence of, for example, teachers play-ing an important role in personally providing material and financial support to impoverished families when they fail to access other services.

It is therefore important that research on policy engages with the differential geographies of policy implementation, rather than producing 'critiques' of policy in a blanket way. Indeed there is a vein of research around family policy in parti-cular (Furedi, 2001) which implicitly or explicitly undermines any role of the state or workers in intervening in families' lives, through a blanket critical stance on government policy's concerns with personal life. This approach potentially overlooks the basic fact that poor families do need state support, particularly in the current economic climate. I would therefore like to suggest the need for a different tone of research, which sits alongside critical deconstruction but also draws on the potentialities of emotions around encounters between state and families, through being based on the complexities, 'messiness' and felt experience of practice and also of living in a poor and vulnerable family. Whilst the expe-riences of poor families are offered up within policy and media discourses in stigmatising ways that, as Levitas (2014) comments, 'evoke a burgeoning dys-functional underclass' and more official policy orientated 'evidence' relies on blanket categories of families and needs, an empirical approach is needed which allows for the experiences of front-line workers and the situations of poor families to be explored in a more empirically attentive and compassionate way. In a similar vein Featherstone et al. (2014: 153) write that social work should be:

> [c]ompassionate and empirical. Social workers must attend to furthering their understandings of the particular family and individuals before them, rather than glossing families into spurious universals and institutional categories.

In the final part of this chapter I will discuss a number of concepts drawn from feminist theory, and particularly debates in feminist care ethics, which I suggest are productive for developing such an empirical approach for both professionals and researchers. This section therefore seeks to suggest a new 'structure of feeling' or emotional repertoire around family policy, which might usefully shape both public discourse and everyday encounters. Therefore the intention here is not simply to pit an empirical attentiveness to everyday practices against stigmatising wider public discourses, but rather to consider how such an empirical attentiveness might itself become part of a wider political project and discourse.

Care, vulnerability and emotional resources

What broader understandings of the (emotional) self lie behind some of the affective patternings in policy and popular culture around poor families discussed in previous sections? A key issue here is dependence and independence. To return to the quotation from the Children's Centre research about the child who had fallen over, there is a sense of an affective shift at work away from dependence and care towards an ideal of independent families who do not require state support, either at the level of the prosaic everyday interactions of picking up a crying child, or at the level of the kinds of financial state benefits now being reduced. Current UK political rhetoric is saturated with imaginings of hard-working and resourceful families and the need to do away with 'dependence' in all its forms (Cameron, 2011). However, as also noted, such imperatives are often resisted and challenged at the level of everyday practice, for example in Children's Centres.

Feminist care ethics 'fundamentally challenges the idea of the autonomous individual' (Ribbens McCarthy, 2014: 344), instead placing 'vulnerability and dependence at the centre of human experience' and the need therefore to value practices and processes of care expressed through emotional dynamics of 'attentiveness, responsibility, trust, responsiveness' (Ribbens McCarthy, 2014: 344). Schues (2014) goes beyond simply recognising vulnerability but insists it is a prerequisite to all engagement with the world, including being political, giving birth and making friendships. Such vulnerability would therefore encompass practices and experiences of professionals as well as service users. Of course, it is not sufficient to pay attention to vulnerability and care without attending to the specific empirical sets of conditions and power relations which surround both the giving and receiving of care and the production of vulnerability in different forms. Schues and others (see Brugere, 2014) talk of the need for 'an anthropology of vulnerability'. In relation to disadvantaged families, such an empirical project might encompass the geographies as well as temporalities of vulnerability and disadvantage, in ways that demonstrate how families can move in and out of being 'troubled' or experiencing troubles, rather than inhabiting such a position or label forever. Indeed, the importance of understanding family change over time is a key theme in Ribbens McCarthy et al.'s (2014) collection, which seeks to challenge the categorisations of 'normal' or 'troubled' families. In a similar vein, Sophie Bowlby and others (2010) have employed the notion of 'caringscapes' as a way to understand the 'landscapes' of care of families over time, encompassing both the resources and vulnerabilities they encounter and the pathways through which they navigate them. This project might therefore usefully draw on the approaches of human geography to understand poor families as emplaced within particular contexts and communities (see Evans and Holland, 2012), and to understand what kind of family support work is needed to attend to these wider contexts.

This empirical approach could also contribute to a different kind of emotionally imbued wider affective patterning or 'structure of feeling' around poor families, which is rooted in a more compassionate apprehension of families as 'concrete others' (Benhabib, 1985; Varley, 2013). This phrase has been used to discuss the ethical imperative to connect with others across difference without projecting a generalised 'otherness' onto them. Varley (2013) discusses this with reference to understanding female-headed households within development contexts: the need for a detailed empirical attention to the situations and needs of specific families, in their complexity and messiness. This is similar to the call for a 'compassionate empiricism' (Featherstone et al., 2013) quoted above, an approach grounded in a much more nuanced understanding of the felt experiences and material realities of life for poor families. Such an approach, based on care ethics, invokes more quietly emotional (see Askins, 2014) or restrained qualities of reflection, patience and responsiveness rather than the more visceral emotions of pity, shame or disgust evoked, for example, by the documentaries on poverty discussed above.

Conclusion

This chapter, therefore, has considered emotions surrounding family policy and professional practice in a number of ways. I have discussed 'structures of feeling' and 'affective patterning' at the level of policy and popular culture. I have argued that these wider emotional currents inevitably shape practice on the ground but are also resisted and reworked within encounters between families and professionals. I have also begun to sketch out ways in which a more 'concrete' understanding of poor families' experiences could be used to inform a wider politics and ethics, drawing on a feminist ethic of care and an exploration of the wider geographies and temporalities that shape family lives.

To conclude, I want to return briefly to the incident involving myself and my son discussed earlier. What kinds of affective and emotional cues led the health visitor to judge me and the other family in such different ways? What might a more reflective and compassionate approach have yielded? In relation to myself, she failed to recognise my own vulnerability and need for support. Much more importantly, her view of the other family as 'in need' involved moral judgement and a stigmatising (semi-public) labelling of them. A differently imbued view might have considered the specific circumstances in which they found themselves, their 'anthropology of vulnerability' and the resources that might enable them to tackle the problems they were experiencing, as well as a recognition that these problems did not define them as a family, that they were not permanent. This was how she understood my family, after all.

However, this chapter seeks to go beyond suggesting particular kinds of professional practice and the emotional competences involved. To return to the points with which I opened this chapter, it is important to develop accounts of emotional governance that connect the broader emotional patternings of policy and media to analysis of specific interactions between professionals and service users and the particular geographies of care within which poor families are situated. This is because not only are the feelings of both professionals and service users shaped by these wider patternings, but also because affective practices and the specificities of experiences can contest these public moods. Indeed, beyond exploring such concrete practices, there is an urgent need to develop an ethics and politics of emotion that suggest progressive alternatives to current dominant emotional political landscapes. While this chapter has focused on the stigmatis-ation attached to poor families at the current time in the UK, a similar analysis might be undertaken in relation to how feelings about migrants and people with disabilities are being shaped across the emotional landscapes of policy and media. Research that seeks to counter these negative feelings might connect with different emotional registers, of compassion, care and reflection, qualities without which there can be no progressive welfare settlement.

Notes

1 Health visitors are registered nurses with a specialist qualification in community health and health promotion for families with young children.
2 The emotional politics of the 'Baby P' case were particularly concerned with the failure of social workers and NHS workers to intervene, despite repeated contact with the family.
3 The research was carried out between 2009 and 2011, before the full impacts of the policies of the current Conservative government (2010 to present) were being realised in such settings.
4 All names of places and participants are pseudonyms. More details of the research methodology can be found in Jupp (2012) and Jupp (2013a).

References

All website URLs were accessed on 16 April 2015.

Ahmed, S. (2004) *The cultural politics of emotion.* Routledge, New York.

Allen, K., Tyler, I. and De Benedictus, S. (2014) 'Thinking with "White Dee": the gender politics of "austerity porn"', *Sociological Research Online*, 19(3): 2.

Askins, K. (2014) 'A quiet politics of being together: Miriam and Rose', *Area*, 46(4): 353–54.

Bailey, N. (2012) 'Policy based on unethical research', *Poverty and Social Exclusion Online*, at: www.poverty.ac.uk/news-and-views/articles/policy-built-unethical-research.

Barnes, M. and Prior, D. (eds) (2009) *Subversive citizens: power, agency and resistance in public services*. Policy Press, Bristol.

Benhabib, S. (1985) 'The generalized and the concrete other', *Praxis International*, 4: 402–24.

Bowlby, S., McKie, L., Gregory, S. and Macpherson, I. (2010) *Interdependency and care over the lifecourse*. Routledge, London.

Brugere, F. (2014) 'Emotions as constituents for an ethic of care', paper given at the Oxford Institute of Population Ageing Seminar series, 29 October 2014, at: www.oxfordmartin. ox.ac.uk/event/1963.

Cameron, D. (2011) 'Fightback after the riots', speech available at: www.gov.uk/government/ speeches/pms-speech-on-the-fightback-after-the-riots.

Clarke, J. and Newman, J. (2012) 'The alchemy of austerity', *Critical Social Policy*, 32(3): 299–319.

Clayton, J., Donovan, C. and Merchant, J. (2015a) 'Emotions of austerity: care and commitment in public service delivery in the north east of England', *Emotion, Space and Society*, 14: 24–32.

Clayton, J., Donovan, C. and Merchant, J. (2015b) 'Distancing and limited resourcefulness: third sector provision under austerity localism in the north east of England', *Urban Studies*, published online 21 January 2015, doi: 10.1177/0042098014566369.

DCLG (Department of Communities and Local Government) (2012) *Listening to troubled families: a report by Louise Casey*. HMSO, London.

DCLG (Department of Communities and Local Government) (2014) 'Troubled Families programme turning 117,000 lives around', at: www.gov.uk/government/news/troubled-families-programme-turning-117000-lives-around.

Dermott E. (2012) 'Poverty vs parenting, an emergent dichotomy', *Studies in the Maternal*, 4(2), at: www.mamsie.bbk.ac.uk/back_issues/4_2/index.html.

England, K.V.L. (1994) 'Getting personal: reflexivity, positionality, and feminist research', *Professional Geographer*, 46: 80–9.

Evans, R. and Holland, S. (2012) 'Community parenting and the informal safeguarding of children at neighbourhood level', *Families, Relationships and Societies*, 1(2): 173–90.

Featherstone, B., Morris, K. and White, S. (2013) 'A marriage made in hell: early intervention meets child protection', *British Journal of Social Work*, 44: 1735–49.

Featherstone, B., White, S. and Morris, K. (2014) *Re-imagining child protection: towards humane social work with families*. Policy Press, Bristol.

Field, F. (2003) *Neighbours from hell: the politics of behaviour*. Politico's Publishing, London.

Furedi, F. (2001) *Paranoid parenting*, Bloomsbury, London.

Gambles, R. (2010) 'Going public? Articulations of the personal and political on Mumsnet. com', in Mahony, N., Newman, J. and Barnett, C. (eds) *Rethinking the public*. Policy Press, Bristol, pp. 75–90.

Gerhardt, S., Jowell, T. and Stewart-Brown, S. (2011) 'You don't talk about love in government', *Soundings*, 48, Summer: 145–57.

Gillies, V. (2007) *Marginalised mothers: exploring working-class experiences of parenting*. Routledge, London.

Jensen, T. (2014) 'Welfare commonsense, poverty porn and doxosophy', *Sociological Research Online*, 19(3): 3.

Jensen, T. and Tyler, I. (2012) 'Editorial. Austerity parenting: new economies of parent-citizenship', *Studies in the Maternal*, 4(2), at: www.mamsie.bbk.ac.uk/back_issues. html.

Jones, H. (2013) *Negotiating cohesion, inequality and change: uncomfortable positions in local government.* Policy Press, Bristol.

Jupp, E. (2010) 'Public and private on the housing estate: small community groups, activism and local officials', in Mahony, N., Newman, J. and Barnett, C. (eds) *Rethinking the public.* Policy Press, Bristol, pp. 75–90.

Jupp, E. (2012) 'Parenting policy and the geographies of friendship: encounters in a Sure Start Children's Centre', in Kraftl, P. and Horton, J. (eds) *Critical geographies of childhood and youth.* Policy Press, Bristol, pp. 215–29.

Jupp, E. (2013a) '"I feel more at home here than in my own community": approaching the emotional geographies of neighbourhood policy', *Critical Social Policy*, 33(3): 532–53.

Jupp, E. (2013b) 'Enacting parenting policy? The hybrid spaces of a Sure Start Children's Centre', *Children's Geographies*, 11(2): 173–87.

Jupp, E. (2016) 'Families, policy and place in times of austerity', *Area*, available at Online First.

Levitas, R. (2014) '"Troubled Families" in a spin', *Poverty and Social Exclusion Online*, at: www.poverty.ac.uk/editorial/%E2%80%98troubled-families%E2%80%99-spin.

Lister, R. (2003) 'Investing in the citizen-workers of the future: transformations in citizenship and the state under New Labour', *Social Policy and Administration*, 37(5): 427–43.

Lupton, R. (2003) *Poverty street: the dynamics of neighbourhood decline and renewal.* Policy Press, Bristol.

Pile, S. (2010) 'Emotions and affect in recent human geography', *Transactions of the Institute of British Geographers*, 35: 5–20.

Ribbens McCarthy, J. (2014) 'What is at stake in family troubles? Existential issues and value frameworks', in Ribbens McCarthy, J., Hooper, C.-A. and Gillies, V. (eds) *Family troubles.* Policy Press, Bristol, pp. 327–53.

Ribbens McCarthy, J., Hooper, C.-A. and Gillies, V. (eds) (2014) *Family troubles: exploring changes and challenges in the family lives of children and young people.* Policy Press, Bristol.

Rose, N. (1989) *Governing the soul: the shaping of the private self.* Free Association Books, London.

Rutter, M. (2014) 'The role of science in understanding family troubles', in Ribbens McCarthy, J., Hooper, C.-A. and Gillies, V. (eds) *Family troubles.* Policy Press, Bristol, pp. 45–58.

Schues, C. (2014) 'Bodily and social vulnerability: a phenomenological perspective on the practice of care', paper given at Oxford Institute of Population Ageing Lecture Series, 26 November 2014, at: www.oxfordmartin.ox.ac.uk/event/1967.

Tickle, L. (2014) 'Food, clothes, transport, beds, ovens: the aid schools are giving UK pupils', *The Guardian*, 14 October 2014, at: www.theguardian.com/education/2014/oct/14/schools-providing-basic-necessities-to-disadvantaged-pupils.

Varley, A. (2013) 'Feminist perspectives on urban poverty', in Peake, L. and Rieker, M. (eds) *Rethinking feminist interventions into the urban.* Routledge, London, pp. 125–41.

Warner, J. (2013) 'Social work, class politics and risk in the moral panic over Baby P', *Health, Risk and Society*, 15(3): 217–23.

Warner, J. (2014) '"Heads must roll"? Emotional politics, the press and the death of Baby P', *British Journal of Social Work*, 44: 1637–53.

Wetherell, M. (2012) *Affect and emotion: a new social science understanding.* Sage, London.

Part IV

Emotions of citizenship and participation

11 The role of multicultural fantasies in the enactment of the state

The English National Health Service (NHS) as an affective formation

Shona Hunter

No one says they want to get rid of the NHS. Everyone praises it across all parties. It's about as powerful a symbol of goodness that we have so it would be too dangerous not to. But for decades now there has nevertheless been a systematic undermining of its core values. This, this is beyond party politics, the Labour government arguably did as much to damage the NHS as any Tory or coalition government. This is about who we want to be as a nation and what we believe is worth fighting for. Too many people have given too much and fought too hard for us to give away what they achieved and to be left with so very little. To those across the whole political party spectrum and to anyone in any position of power or authority I ask you to search your heart and look at who and what you serve. To those who have discarded all principles save that of profit above all else, to those who have turned their backs on the idea of a truly democratic society and aligned themselves to nothing but self interest, to those who have betrayed the vision of equality and justice and compassion for all that – that vision that provided the crucible from which came forth the National Health Service. I say to you . . . You have besmirched the name of Britain; you have made us ashamed of the things of which formerly we were proud; you have offended against every principle of decency and there is only one way in which you can even begin to restore your tarnished reputation: Get out! Get out! Get out!

(Michael Sheen,[1] People's March for the NHS,
1 March 2015, Tredegar, Wales)

In this chapter I consider the role of multicultural institutional fantasies, such as the one I argue can be seen at play in Michael Sheen's speech, in enacting the state. To make such an analysis I build on ideas about the interconnection between socio-cultural relations of power and emotion (Ahmed, 2004; Fortier, 2008; Hage, 1998; Rose, 1996) which can be applied to our understanding of institutions as constituted through 'relational politics' (Hunter, 2012, 2015). Such an analysis can complement and enrich understandings of institutions as produced through political and economic interests by drawing attention to the ways in which emotion works through culture as the connective tissue of institutional life. That is to say that emotion is *productive*, the means by which everyday technical, bureaucratic and professional practices are lived through broader culturally enacted discourses to constitute institutional spaces. From this point of view institutions are not things in themselves, but come into being as 'affective formations', sustained through and sustaining of intersecting material and symbolic relations of classed, racialised and gendered power. Institutions have a fantasy quality, bringing together multiple symbolic and material practices in a particular institutional form. This approach to institutions as affective formations produced through fantasy is not one which denies institutional materialities, but one which facilitates understanding of how the material and symbolic dimensions of life connect through emotion to create particular institutional formations. Thinking about institutions in this way, as material, cultural *and* affective, provides a way of analysing the strength, but also the fragile, complex, uneven and internally contradictory character, of public, professional and political support for institutions like the English NHS.[2] It provides an approach which can explain the existence of cultural continuity amidst ceaseless hyperactive and turbulent material organisational change.

I begin by situating the NHS as constituted through a feel-good fantasy of the nation as solidaristic and equal, a liberal fantasy of equality which is rooted in post-war ideals of material and social emancipation, a fantasy now constructed as under threat in the context of increasing marketisation. In this section I demonstrate how the NHS becomes an object in itself through the multiple investments in it as a symbol of national equality. It is not a representation of the people but an enactment (Mol, 2002) of them, bringing 'the people' into being as a particular sort of collectively established object, a modern equal nation. From the starting point of the NHS as a fantasy which brings the people into collective being around the idea of universal equality, I argue that the strength of controversy over NHS change relates to the destabilisation of this enactment of the people as equal. This is so controversial because it threatens to expose the idealised nature of this ontological fantasy of the nation, its relation to the 'common good' and the hidden inequalities on which it depends. It threatens to expose the profoundly unequal nature of the initial post-war social settlement through which the NHS was founded. This was a settlement between capital and labour which was rooted in British Imperialisms, the problematic exclusions of colonial conquest and violence constitutive of the NHS's material as well as symbolic relations. I then move on to analyse the relational politics of pride and shame which form the connective tissue of this fantasy structure around the 'common good' as part of a newer more complicated multicultural national fantasy which, whilst appearing to be more differentially inclusive than its more liberal democratic predecessor, in fact protects against a related set of (post)colonial[3] anxieties, ones which deepen in the context of austerity politics and dwindling financial resources. This newer multicultural fantasy is part of what blocks more complex conversations around the reproduction of institutional inequality than are allowed for in debates over the respective merits of the market, state or civil society in effective health care provision.

The (post)colonial NHS

Founded in 1948 by the Labour politician Aneurin Bevan and viewed as the capstone of the post-war liberal welfare state, the NHS is unique globally in terms of its stated aims to provide universal health care for all, in response to need, free at the point of use and paid for collectively through general taxation. It was established as a core plank of a broader British post-war commitment to nationalised, state-managed redistributive public services. Of all the institutions designed to tackle the post-war 'five giants' – the social ills of poverty, disease, poor education, squalid housing and unemployment – the NHS is the most clearly locatable in socially solidaristic forms of liberal welfare. Yet from its inception these principles of solidarity have in practice been strongly contested from a range of points of view, professional, political and public (Timmins, 1995). As Sheen highlights in his speech, successive service 'modernisations' have sought organisational, financial and cultural reforms, the general drift of which has been away from state organised and funded provision to more managerially focused arm's length, internally competitive forms of institutionally fragmented provider-driven services, geared towards individual health care needs, patient choice and flexibility (Fotaki, 2006; Speed and Gabe, 2013; Ruane, 2012; West, 2013). Yet despite its shifting and contested nature, the popular common-sense representation of the NHS is remarkably stable. It continues to be positioned, in terms similar to Sheen's points above, as a triumph of liberal social principles of equality and universalism and the best possible expression of the collective national character. The NHS which is so valued is the 'original' one, seen to 'come forth' from a specifically British identity understood through characteristics of decency, goodness and compassion.

However, within the British (post)colonial context ideas of race and nation are conflated, discursively constitutive of a particular intersectional configuration of normative cultural (classed/gendered/generational) relations through which 'the people' are imagined as timeless, homogeneous and stable. This timeless homogeneity and continuity is maintained through shifting negotiations of inclusion within the nation. Taking the example of Sheen's terms: 'This is about who we want to be as a nation and what we believe is worth fighting for.' The continual negotiation of the limits to inclusion rely on the production of various minoritised Others to be included *within* the national culture. Tolerance of difference and contest over its limits are therefore the means by which multicultural inclusion is negotiated (Lewis, 2005). The idea of fantasy is useful because of the way it highlights the nature of discourse which functions to conflate meaning so that one idea operates as a 'master signifier', standing in for multiple sets of ideas and practices gathered together within a network of signification and where discourse operates as a form of collectively held unconscious ideal which enacts meaning in codified rather than explicit ways, highlighting some ideas and obfuscating others. Within a (post)colonial context such as this, the unspoken 'master signifier' within contemporary multicultural fantasies of inclusion is whiteness which functions as a form of 'white nation fantasy' (Hage, 1998) rooted in white supremacist ideals of omnipotence and control over the nation. Within the context of a white nation fantasy, liberal imaginaries of benevolence, inclusion and tolerance of difference within the polity rest on a solidification of difference *outside* its boundaries, but also on differential inclusion *within* it. Thus multiculturalism functions as a form of symbolic violence in which 'a mode of domination is presented as a form of egalitarianism' (Hage, 1998: 87).

In this chapter I am interested in how these ideas of whiteness are enacted in a codified way through institutional discourses of modernisation and the way this codification operates through fantasy (via pride and shame) to enact present-day liberal multicultural institutional space by way of contested continuities with the liberal universalism of the post-war welfare state. I therefore trace how institutional ideals rest on a particular fantasy of equality which, when interpreted through Britain's temporal and spatial positioning within global imperial flows, relies as much on the tolerance and inclusion of some minoritised subjects, as 'ethnic minorities', as it (did and still) does on the exclusion of Others as 'foreigners', migrants, refugees or asylum seekers (Brown, 2006).

Relational politics

The relational analysis of cultural continuity developed in this chapter takes debate around the inequalities enacted through the state away from technical discussions over the merits of bigger versus smaller state, collective versus individual, more bureaucratic versus market or civil society forms of organising, which tend to be associated with *either* political right or left. Instead my relational critique rests on understanding *how cultural power works* through these forms of organising, whether affiliated with the left or right. What is at stake is recognition of how the notion of culture works differently but nevertheless problematically from both left and right. The individualisms which tend to be favoured in right-leaning politics tend to ignore or downplay the role of culture in action. Meanwhile, the collectivisms favoured in the left liberal tradition tend to work in practice to fix culture as a means to highlight common experience. However, both rely on ideas of individual or collective identity as *a priori* things in themselves and as the starting point for action. In contrast, from a relational point of view 'things', like identity (whether understood as individual or collective), are produced through action, enacted differently *through* culture which *connects* individuals to the collective. From this point of view no one unifying ideal, whether framed in terms of individual preference, via ideas of personalisation or patient choice for example, or collective experience, like that posited through ideas of universalism, is *necessarily* more or less productive of equality. All forms of governing order, whether more or less collectively inclined, enact relations of cultural power and inequality.

This relational analysis poses complicated questions for democratic engagement, responsibility and inclusion which mean that no single set of people, politicians, users, professionals or otherwise is ever straightforwardly included or excluded, or straightforwardly innocent or to blame for the unequal ordering of institutional space. The uneven patterning of institutional practices cannot be resolved through different technical manifestations of the state as closer to 'the people', as Sheen's comments imply. Instead, understanding emotion and affect as *constitutive* of (governing) cultures is a starting point for resistance, challenging hierarchically unequal institutional cultures. This potentially offers hopeful prospects for less oppressive, more relationally (rather than liberally) democratically engaged governing cultures.

Such an analysis of the state which understands emotion as constitutive of culture is suggestive of how left-oriented political critiques of the inequalities produced through market individualism can become crucial to the re-production of those very inequalities which form their target. This is because demands for inclusion embedded within such critiques rely on a tightening rather than a loosening of what are in fact complex and contested, culturally enacted links between signification and practice through the creation of an Other(s) which is to be included. Relational analysis provides an important explanation as to *why*, within a context of widespread contestation over the inequalities produced through successive NHS modernisations, the policy direction continues to march marketwise. The *how* is what I am interested in drawing out.

From the point of view of relational politics (Hunter, 2015) it is the connection between feeling and judgements relating to particular objects that is suggestive of the *psychosocial* relationship between politics and the emotions and which points to how fantasy fuels politics and politics fuels fantasy in complex and dynamic ways through relations of difference rather than sameness. Feelings and, especially, our anxieties about who we are frame our judgements of value and our investments in ideas such as equality, or a social category such as race or a set of ideas and practices, such as an institution like the NHS. This 'feeling work' (Gunaratnam and Lewis, 2001) is relational in the sense that it is the means by which the individual and the social are connected (Harding and Pribram, 2002: 424) and culture is enacted *through* the universal experience of difference rather than on the basis of prior identification of ontological similarity. Cultural difference does not form the basis for the identification of difference through the same identity but rather of similarity in experience because of shared *positioning* (through the relations of power) within a social context. *Feeling work* is what produces identification, rather than static identity preceding feeling.

Following Sara Ahmed (2004: 44–7), one way of conceptualising feeling work in dynamic terms is as an affective economy where emotion works as a form of capital in the Marxist sense, gathering value as a function of its circulation. Feeling is understood as the means by which this circulation occurs. Therefore, feeling, such as love and pride, does not reside in an object or a sign such as race or the NHS, but is produced as *an effect* of the circulation between objects (like particular bodies) and signs (like the idea of whiteness or Blackness). Thus supportive attachment to the sign of the NHS, for example we might say love, is an effect of the movement of the idea of the NHS across subject-objects, professions (such as nurses, doctors, physiotherapists), users (such as mother, child, grandparent), politicians, civil servants, pharmaceutical companies, etc. Attachments intensify through love's circulation, generating their own proximities whereby these subject-objects are brought into relationship through their attachment to the symbol of the NHS. Thus the various attachments to, or investments in, an idea generate it as an object, creating its topography. In this way emotions are productive of social relations as well as produced through them, and different sorts of investments are interdependent. Emotions attest to the interdependence of multiplicity and singularity contained within the enactment of the social.

For Sara Ahmed (Ahmed, 2004) thinking about emotion as part of a dynamic affective economy 'allows us to address the question of how subjects become invested in particular structures such that their demise is felt as a kind of living death' (p. 12). Emotion tells us something about processes of objectification, for example how the NHS comes into being as a meaningful object through practice, and also about processes of reification, how the boundaries of that object NHS come to be understood as fixed. Building on this idea of emotions as productive of social structures and being more explicit about the psychosocial nature of the processes involved gives us a way of understanding how institutions come to have meaning and value through multiple investments. From a psychosocial perspective we can posit investment as a process of according value to objects which relates to the idealisation of certain objects as good and as associated with positive images of the self. On the other hand, investment serves by implication to devalue other sorts of objects as bad, associated with something other than the self. Investment is therefore a dynamic process of producing inclusion and exclusion *through emotion*.

This way of understanding investment facilitates understanding of how commitments to institutions are investments in our idealised selves. They are investments in an image of how we would like to be rather than who we already are. Their demise is so painfully felt precisely because this represents a loss of hope and *future potential* to be good, decent, democratic and equal. We can see this idealisation in the face of loss in the debate over NHS reform as this is represented in Sheen's speech. An image of equality is idealised, attached to understandings of the British nation, 'the people' as equal, produced through love, pride and solidarity and where the threats to the NHS are understood as threats, 'betrayals' to the nation, anti-democratic through the problematic creation of separation and competition which is understood to produce the sorts of inequality and social division antithetical to ideals of Britishness. As Sheen puts it 'You have besmirched the name of Britain; you have made us ashamed of the things of which formerly we were proud; you have offended against every principle of decency.' British identity is not some fixed static state driving political action through the experience of similarity. It is the promise associated with a set of ideals *attached to* Britishness, the fantasy of identification around an ideal that drives collective politics *through the impossibility of difference* via the emotions of pride and shame.

Methodologically my feminist psychosocial approach relies on understanding the contested, relationally enacted nature of meaning which necessitates an analysis of 'evidence of things not seen' (Lewis, 2000: chapter 5 title). Rooted in the idea of the discursive unconscious as a form of temporal and spatial displacement which hides complex relational connections, it aims to expose and re-create links within a discursive network of signification. Expressions of emotion signal connections, relations of proximity and distance which require unpacking in order to interpret meaning hidden within textual expression (whether written, verbal or visual). But reading for emotion is not straightforward. Emotions do not correlate directly to their expression. Emotions flag up contradiction between material practices and processes of signification. They highlight significance, but substantive meaning and impact need to be understood from within the relational context. The aim of such an analysis is therefore not to read for the truth of emotion, but to understand its function within the enactment of meaning. At the level of institutional discourse *alone*, which is the focus in this chapter, this form of analysis *can only ever be suggestive*. This is precisely because fantasy and practice do not align. What such a discursive analysis necessitates in terms of my concept of relational politics and what is missing from my analysis here (but which I explore elsewhere in relation to this example of the NHS, see Hunter, 2009, 2010, 2015) is how policy discourses are enlivened, negotiated, resisted and reproduced. It is through these everyday relational dynamics that fantasies are potentially enacted differently, offering the prospect of institutional change in the context of the apparent continuity that I analyse here.

Our cherished NHS

Contestation over public service reform in the English context has rarely been more pronounced than in relation to the most recent changes to the NHS heralded by the enactment of the 2012 Health and Social Care Act. The explicitly emotional nature of the debate and the strength of defensive protest over the Act are striking (Ashcroft, 2015). It created the first and most obvious split in opinion within the 2010–15 Conservative–Liberal Coalition Government, with the Liberal Democratic Deputy Leader of the Coalition pronouncing that 'protecting the NHS rather than undermining it is now my number one priority'.[4] The Act was denounced by key professional groups such as the British Medical Association, left of centre political groups like the Socialist Party and the Trades Unions, and right of centre and independent groups such as the Kings Fund and the Nuffield Trust. Controversy was so extensive and widespread that 2011 saw an almost unprecedented halt in the official legislative process for a government 'listening exercise' where the Conservative Prime Minister and Secretary of State for Health would 'take time to try to find compromise amendments' (Glynos et al., 2014: 53–6). Resistance to the Act generated new public protests such as those exemplified in the 'People's March for the NHS' established by a group of mothers from Darlington in the north of England.[5] It was as part of this ongoing campaign that Michael Sheen delivered the speech cited at the beginning of the chapter.

The main thrust of this range of criticism was around the Act's shift to a more advanced form of NHS privatisation. The critics' worst-case scenario predictions envisage the establishment of a two-tier system of health care, building on successive reforms of previous Labour and Conservative led governments. Some more controversial aspects of the Act include moves to introduce price regulation as a mechanism for increased competition for cheaper forms of service provision; the lifting of the cap on private provision via the 'any qualified provider' mechanism which would enable the extension of private finance via large corporate health providers, effectively externalising the already internalised market mechanism; the abolition of the remaining two regional planning mechanisms, Strategic Health Authorities and Primary Care Trusts, and the introduction of General Practitioner consortia in their place. Criticism anticipated that together these moves would produce downward pressure on pricing with less regulation for health outcomes and increased incentivisation to grow private provision. The government's justification for such change is the context of financial deficit and the future affordability of the NHS.

The tenor of much of the criticism is reflected in the analysis of Colin Leys and Stewart Player who say:

> [The NHS's] founding principles of comprehensiveness and equal access for all have been core values of modern British society. Working to marketize it, and finally privatizing it, without any democratic mandate . . . is as close as it gets to being . . . unconstitutional. The question is whether the English people . . . will accept having this precious part of our heritage filched from under our noses.
>
> (Leys and Player, 2011: 10–11)

The couching of the critique of marketisation within a call to modern British values, ideas of heritage and 'the people' is interesting. As I have been arguing and many commentators themselves accept, *in bureaucratic practice* the NHS has for some time borne relatively little resemblance to that established by Bevan in 1948. This very obvious contradiction between the ideal and reality raises the question as to what is it precisely that is happening here, and how and why it is that the idea of the NHS engenders such strong support in such a diverse and not necessarily obviously aligned range of professionals, activists and the 'general' public. It is precisely this contradiction between the idea and the material practices of the NHS that alerts us to its fantasy nature, so the question becomes: what is the nature of the fantasy and *what does it elide within* this idea of Britishness and 'the people'?

Re-visioning national pride through tolerance

For Bevan the post-war NHS was:

> [T]he biggest single experimentation in social service that the world has ever seen undertaken. It is, I think, a great tribute to the vitality and genius of the British people that we are able to undertake a task of this complexity and

magnitude within three years of a great war. It shows that the British people have still got the principles of innovation, and renovation, running through them yet, and that we can pioneer in many directions for the rest of the world to follow.

(Bevan cited in Webster, 1991: 140)

From this point of view social service is fundamental to Britishness, a key component of Britain's claims to post-war world leadership. The imperialist tone of Bevan's vision is made more explicit in William Beveridge's view as to the nationalist imperialist justification for the sort of post-war institutional reconstructive experiment as exemplified by the NHS.

Pride of race is a reality for the British as for other peoples . . . as in Britain today we look back with pride and gratitude to our ancestors, look back as a nation or as individuals two hundred years and more to the generations illuminated by Marlborough or Cromwell or Drake,[6] are we not bound also to look forward, to plan society now so that there may be no lack of men or women of quality of those early days, of the best of our breed, two hundred and three hundred years hence?

(Beveridge cited in Virdee and Cole, 2000: 42)

Similar to the contemporary discourse in defence of the NHS, pride works very explicitly in Beveridge's comments as a driver for collective investment in the idea of the NHS. It connects collective equality to the social service inherent in Britishness. In the immediate post-war aftermath pride functions paradoxically as a means to connect 'the people' through ideas of lineage, kinship and raced structure established through 'successful' imperial conquest which is presented as crucial to maintaining the British way of life understood through a commitment to social service. Pride circulates, generating its own momentum. Pride in race generates pride in the institutional structures of the NHS, which in turn sustains pride in the British race as a benevolent, quality people. Despite the continuities across these contemporary defences of the NHS and its post-war impetus, there are important discursive shifts between then and now. These mark shifts away from the explicit representation of the NHS within this broader imperial endeavour in which Britain positioned itself as a post-war leader.

Since that post-war point there has been a break-up of this discursive consensus around the nature and value of the welfare state from social movements on the political left as well as from the right's perspective of market individualism. A postcolonial strand of the left critique highlights the imperialist nature of this enactment of the welfare state, and the NHS in particular, as the culmination of the settlement between capital and labour in which increasing numbers of working class people, and also women (amongst the range of the nation's Others), were incorporated into the nation via the social rights of welfare citizenship (Ahmad, 1993; Cohen, 1996; Doyal and Pennell, 1994; Williams, 1989). From this perspective the development of social welfare and Imperialism are interdependent aspects of the same story. In the British (post)colonial context this critique has taken its cue from the assertive resistance of a range of anti-racist, feminist and other new social protest movements arising from the uneasy settlement of post-colonial Black subjects. Their demands for equality expose the inherent contradictions within this initial post-war settlement as *dependent* on the labour of (largely working-class) women and Black subjects, many of whom migrated through invitation and various 'reciprocal arrangements' with former colonial territories, such as those for the recruitment and training of overseas doctors (Kyriakedes and Virdee, 2003).

These postcolonial critiques have shown the institutionally racist and sexist terms through which such postcolonial settlements are enacted in practice through processes of differential inclusion into state institutions and the provision of a range of welfare services (Lewis, 2000). The sorts of experiences include limited progression and promotion opportunities, the confinement of certain forms of staff to particular specialisms, for example the disproportionately high numbers of south Asian doctors working in the less prestigious discipline of general practice rather than more prestigious hospital contexts and the disproportionately low numbers of minority ethnic staff in managerial and strategic posts. This is at the same time as health care outcomes which straggle far behind those of the rest of the population (Nazroo, 2003; Salway et al., 2010). Whilst it is not clear precisely how the changes in the 2012 Act will impact this situation, the trend since the early to mid-2000s has been a *decrease* in minority ethnic staff in senior management positions, stasis for women staff, continued experiences of discrimination, harassment and bullying for a range of Black and minority ethnic staff and a continued paucity of data on health outcomes for a range of minority ethnic groups in relation to health inequalities (Kline, 2014; Psoinos et al., 2011; Rao, 2014; Turner et al., 2013).

Despite ongoing material and social realities of inequality reproduced through processes of differential inclusion, such processes have also created significant policy reformulations at national and institutional levels. There is now a discursive space in which to recognise, value and even celebrate difference within discourses of equality. From within this discursive reformulation the NHS has been re-visioned at the heart of these (post)colonial processes of social change which have produced a more socially differentiated multi- rather than supposedly mono-cultural national context. For example, Ivan Lewis, Parliamentary Secretary of State for the Department of Health with responsibility for equalities, outlined this reconfigured understanding of equality in a speech commemorating the 60th anniversary of the NHS:

> Coincidentally, [this NHS birthday year] is also the 60th anniversary of the arrival of the SS *Empire Windrush*.[7] So what better moment to pay tribute to the many Caribbean and Asian people who travelled to the UK in the 1940s and subsequent decades and helped build the NHS into the world class service that, in so many ways, it has become today . . . One of the great triumphs of the NHS, largely tax funded, universal and free at the point of need, is that it

is fair and equitable. Indeed, equality was a founding principle of the health service – only by building equality into every aspect of our work will we create a truly person-centred and responsive service.

(Lewis, 2008:5)

There is an interesting shift here in Lewis's comments whereby a more differentiated form of equality and fairness based on the inclusion of Black and minority ethnic staff into the institution of the NHS gets connected with ideas of person-centredness. Pride in ideals of equity remains a constant means to sustain the feel-good fantasy of the NHS as a triumph of the drive to equality in post-war welfare. But in Lewis's comments pride works within the context of changing institutional relations organised around ideas more closely associated with more individualised forms of patient care filtered through principles of personalisation, choice, flexibility and integration popularised within modernisation and market-isation discourses, precisely the sorts of ideas which have been further strengthened within the 2012 Act. Such ideas serve to smooth out, rather than join up, the contradictions between aspects of the staff inequality issues outlined above and positive ethnic minority patient experience (Kaehne, n.d.; Psoinos et al., 2011), and there is little evidence to suggest that these ideas engage with the racialised hierarchies of power which impact on ethnic minority patient health outcomes in terms of access to and experience of health care provision (Salway et al., 2010).

The 2003–5 Labour Health Secretary John Reid and the head of the Commission for Racial Equality Trevor Phillips are even bolder in their representation of the NHS as the pinnacle of British multicultural success, which can reformulate difference through ideals of equality. They claim:

What the NHS as a living and giving organisation tells us is that all this apparent foreignness, all these different others living and working in our midst, are not others. In fact they are melded together by this British institution into 'us'. It is a British NHS run within British values of equity and tolerance and it encompasses all of this diversity within its Britishness. Just as in 1948 the NHS showed us the best way to live with each other, so the NHS in 2004 shows how a nation based on hundreds of different cultures can work together for the good of us all.

(Reid and Phillips, 2004: 2)

For Reid and Phillips the NHS is very clearly constituted through the relations of ethnic difference. From this point of view difference has a positive value. Yet, there is a familiar set of integrationary power dynamics at play here (Raghuram, 2007). Control and agency lie with the British 'host' institution into which 'others' are melded. This is a familiar revisionist discourse which serves to reinforce ideas of benevolence through tolerance on the part of the British nation. In this particular example the NHS is positioned through its integrative function which accords it a key role in re-establishing the nation as equal and for equality within the context of continued antagonism around difference and inequality. Pride shifts its meaning to encompass diversity as crucial to visions of national equality, but in a way which bolsters revisionist narratives of the British nation. It produces a 'feel good' (Ahmed, 2008) fantasy of the nation as socially cohesive and equal.

The political logic of the proud feel-good fantasy remains consistent, even as its substantive content shifts. It remains revisionist in nature in the sense that collective benevolence is understood as remaining at the heart of the nation. This fantasy of revisionism is what guards against developing a more realistic view of ongoing institutionalised inequality as fundamental to the nation. Threats to the revisionist narrative placing the NHS in this position as symbol of national solidarity are more than threats to health-care provision as such. They threaten exposure of the differentiated and multiple nature of the polity itself through the exposure of broader racist imperialist historical material relations through which its institutions come to be. Given the roots of the NHS in left liberal social democratic principles, this postcolonial critique exposes the fissures in the left's fantasy of equality and its image of itself as a force for good.

The contradiction of the march towards marketisation and ongoing commitments to equality is explained through the feel-good fantasy producing the NHS. This fantasy is threatened by the recognition of difference as a cause of fragmentation. Ironically, threats to the validity of the NHS ideal itself produced through the recognition of difference are domesticated and reduced through ideals of marketisation such as choice personalisation and responsiveness, and of integration (West, 2013). This is precisely because these ideas work with other ideas such as diversity and difference to create the illusion of social and cultural inclusion. The social democratic left's commitment to ideas of universal equality established through post-war social democratic principles prevent it from recognising the full implications of postcolonial critiques of the *fundamentally constitutive nature of inequality*. It is this dogged commitment that creates the space for marketisation to enter via the appropriation of certain ideas of individualised forms of inclusivity. Thus marketisation enters by stealth through the back door of ideas of personalisation, choice, flexibility as a potential means for *protection* from the reality of a polity which is not cohesive, but which is already divided from itself; characterised by difference, rather than sameness.

Conclusions: for the multicultural love of the (working classed) people

Because of the capacity for fantasy to obfuscate complex and contradictory social realities, an account of affective force must be accompanied by an exploration of the parameters of comprehension of the (political) logic of the fantasy that supports an affective economy and provides its coherence. Analysis of the affective force driving the fantasy logic helps to answer the question as to what function the fantasy is performing. In this case I suggest the affective force of this multicultural fantasy of the NHS serves to cover up what is actually an internally differentiated and complicatedly acrimonious enactment of 'the people', where not everyone is included on the same terms, and even more problematically where certain people's inclusion is predicated on the unequal inclusion of others. In the case of the NHS, working-class people's and women's emancipation was enabled through the labour of others as part of the shifting boundary negotiations around what constitutes tolerance. The overarching fantasy ideal of equality and solidarity represented in the NHS is enacted through multiple potentially different meanings invested in it. However, this overarching multicultural fantasy continues to support a singular meaning of equality which hides the internal multiplicity of points of investment. Different forms of relationality make up what appears to be a coherent institutional whole, but which is in fact internally and hierarchically differentiated. Fantasy hides certain meanings and relations and brings others to the fore. It therefore hides the relational politics which go into its production, creating itself as a reality. The contemporary multicultural fantasy constitutive of the NHS provides us with an example of the reification of discourses through the elision of an understanding of the complicated dynamics of power.

My point here is not that fantasy itself is bad or necessarily problematic. On the contrary, it is fundamental to the construction of a liveable social world, crucial to symbolisation and meaning production which make collective action possible. On the one hand, fantasy is enabling, and the fantasy of equality has certainly been so in bringing the NHS into being as an enduring symbol of solidarity which has produced much social good through the reduction of health inequality. Even strident critics of post-war gendered Imperialism agree (Doyal and Pennel, 1994). Fantasy is helpful in bringing together very different feelings and ideas about the NHS, containing difference within it, producing a means of working together, a point of solidarity *through difference*. This fantasy *can* function as a container for social difference, providing collective responsibility and care for the Other. And there are certainly arguments that post-war welfare ideas of universalism and collectivism went a long way to producing this sort of important container (Hoggett, 2000). Instead what I suggest is that it is the point at which fantasy hardens, where it becomes a reification of multiple changing complex practices and ideals, the point at which it becomes *normative*, becoming a means to over-simplify and deny complex realities, that it becomes problematic. At this point fantasy becomes delusion (Hoggett, 2000), a point for manic investment in an object as a singular thing in itself, which bears little relation to the multiple complexes which make it up. At this point fantasy becomes repressive, eliding the complex relations of power and inequality through which it is made. Looking back at Sheen's comments, they risk reinforcing this sort of repressive delusional quality encompassed in many of the most vociferous contemporary defences of the NHS, splitting good and bad where there is only one option for change, and the defence of the NHS becomes a manic 'no matter what' defence. Such delusion conflates the differentiated practices making up the NHS with the NHS presented as a particular coherent and continuous sort of thing in itself. This conflation narrows opportunities for institutional change through different practices. This is why it becomes so important to unpick the broader relational politics out of which a particular fantasy and its form prevail. In this case the socio-historical enactment of the NHS and other British welfare institutions through the settlement between working-class labour and capital becomes crucial to understand the force of the ideal of equality and the problematic hardening of such an ideal as being of a particular capitalist imperialist sort. Failing to understand this multiply uneven power dynamic potentially serves to harden, rather than problematise, the complicated power relations whereby the relationship between the whitened (deracialised) British majority and the nation's racialised Others is severed. The outcome of this process of elision is the (potential) production of division and blame which splits rather than connects differentiated others. It undermines the desire for social solidarity which so much effort is being put in to save.

Notes

1 Michael Sheen is a well-known actor. This speech was inspired by Bevan and delivered in his home town on St David's Day.
2 I use English rather than British here as there are differences in how the National Health Service (NHS) is organised in each of the four countries of England, Scotland, Northern Ireland and Wales.
3 I bracket the post in (post)colonial here to problematize the temporal discontinuity suggested through the use of post. My point is to emphasise the continuity between imperial history and present-day institutional practices. However, I do not use brackets where I refer to analytic approaches which define themselves in terms of postcoloniality. This emphasis of continuity is better represented through decolonial critique.
4 Speaking on the Andrew Marr Show, BBC1, 8 May 2011. Available at: www.bbc.co.uk/news/uk-politics-13325975 (accessed 22 June 2016).
5 These mothers were nicknamed the Darlomums. Their march covered 23 towns and cities, following in the steps of the famous 1936 Jarrow marchers during the Great Depression who marched 300 miles from Darlington to Parliament Square in London in order to protest against unemployment and poverty in the north of England.
6 The Duke of Marlborough, Sir Francis Drake and Oliver Cromwell are all significant figures in British military imperial conquest of various parts of the globe.
7 The *Empire Windrush* was the ship which carried one of the first large groups of invited migrants from the Caribbean to undertake post-war reconstruction work in Britain.

References

Ahmad, W.I.U. (1993) 'Making black people sick: 'race', ideology and health in research', in Ahmad, W.I.U (ed.) *Race and health in contemporary Britain*. Open University Press, Milton Keynes.

Ahmed, S. (2004) *The cultural politics of emotion*. Edinburgh University Press, Edinburgh.

Ahmed, S. (2008) 'The politics of good feeling', *Australian Journal of Critical Race and Whiteness Studies Association*, 4(1): 1–18.

Ashcroft, M. (2015) 'The people, the parties and the NHS', *Lord Ashcroft Polls*.

Brown, W. (2006) *Regulating aversion: tolerance in the age of identity and empire*. Princeton University Press, Princeton, NJ.

Cohen, S. (1996) 'Anti-semitism, immigration controls and the welfare state', in Taylor, D. (ed.) *Critical social policy: a reader*. Sage, London.

Doyal, L. and Pennell, I. (1994 [1979]) *The political economy of health*. Pluto Press, London.

Fortier, A.M. (2008) *Multicultural horizons: diversity and the limits of the civil nation*. Routledge, London.

Fotaki, M. (2006) 'Choice is yours: a psychodynamic exploration of health policymaking and its consequences for the English National Health Service', *Human Relations*, 59(12): 1711–44.

Glynos, J., Speed, E. and West, K. (2014) 'Logics of marginalisation in health and social care reform: integration, choice, and provider-blind provision', *Critical Social Policy*, 35(1): 45–68.

Gunaratnam, Y. and Lewis, G. (2001) 'Racialising emotional labour and emotionalising racialised labour: anger, fear and shame in social welfare', *Journal of Social Work Practice*, 15(2): 131–48.

Hage, G. (1998) *White nation: fantasies of white supremacy in a multicultural society.* Pluto Press, Annandale, Australia.

Harding, J. and Pribram, D.E. (2002) 'The power of feeling: locating emotions in culture', *European Journal of Cultural Studies*, 5(4): 407–26.

Hoggett, P. (2000) *Emotional life and the politics of welfare.* Palgrave, London.

Hunter, S. (2009) 'Subversive attachments: gendered, raced and professional relignments in the "new" NHS', in Barnes, M. and Prior, D. (eds.) *Subversive citizens: power, agency and resistance among public service users and workers.* Policy Press, Bristol, pp. 137–53.

Hunter, S. (2010) 'What a white shame: race, gender, and white shame in the relational economy of primary health care organizations in England', *Social Politics: International Studies in Gender, State and Society*, 17(4): 450–76.

Hunter, S. (2012) 'Ordering differentiation: reconfiguring governance as relational politics', *Journal of Psycho-Social Studies*, 6(1): 3–29.

Hunter, S. (2015) *Power, politics and emotions: impossible governance?* Routledge, London.

Kaehne, A. (n.d.) 'One NHS, or many? The National Health Service under devolution', *Political Insight*, at: www.psa.ac.uk/insight-plus/one-nhs-or-many-national-health-service-under-devolution (accessed 6 August 2015).

Kline, R. (2014) *The 'snowy white peaks' of the NHS: a survey of discrimination in governance and leadership and the potential impact on patient care in London and England.* Middlesex University, London.

Kyriakides, C. and Virdee, S. (2003) 'Migrant labour, racism and the British National Health Service', *Ethnicity and Health*, 8(4): 283–305.

Lewis, G. (2000) *'Race', gender, social welfare: encounters in a postcolonial society.* Polity Press, Cambridge.

Lewis, G. (2005) 'Welcome to the margins: diversity, tolerance, and policies of exclusion', *Ethnic and Racial Studies*, 28(3): 536–58.

Lewis, I. (2008) 'Equality in the Health Service: 60 years on', *Diversity in Health and Social Care*, 5: 5–6.

Leys, C. and Player, S. (2011) *The plot against the NHS.* Merlin Press, Wales.

Mol, A. (2002) *The body multiple: ontology in medical practice.* Duke University Press, London.

Nazroo, J. (2003) 'The structuring of ethnic inequalities in health: economic position, racial discrimination, and racism', *American Journal of Public Health*, 93(2): 277–84.

Psoinos, M., Hatzidimitriadou, E., Butler, C. and Barn, R. (2011) *Ethnic monitoring in healthcare services in the UK as a mechanism to address health disparities: a narrative review.* Swan IPI, London.

Raghuram, P. (2007) 'Interrogating the language of integration: the case of internationally recruited nurses', *Journal of Clinical Nursing*, 16(12): 2246–51.

Rao, M. (2014) 'Inequality rife among black and minority ethnic staff in the NHS', *The Guardian*, 1 August 2014, at: www.theguardian.com/healthcare-network/2014/aug/01/inequality-black-ethnic-minority-rife-nhs (accessed 10 March 2015).

Reid, J. and Phillips, T. (2004) *The best intentions? Race, equity and delivering today's NHS*. Fabian Society, London.

Rose, J. (1996) *States of fantasy*. Oxford University Press, Oxford.

Ruane, S. (2012) 'Division and opposition: the Health and Social Care Bill 2011', in Kilkey, M., Ramai, G. and Farnsworth, K. (eds) *Social policy review 24: analysis and debate in social policy 2012*. Policy Press, Bristol, pp. 97–114.

Salway, S., Nazroo, J., Mir, G., Craig, G., Johnson, M. and Gerrish, K. (2010) 'Getting to grips with health inequalities at last?', *British Medical Journal, Rapid Response*, at: www.bmj.com/rapid-response/2011/11/02/fair-society-healthy-lives-missed-opportunity-address-ethnic-inequalities- (accessed 13 August 2015).

Speed, E. and Gabe, J. (2013) 'The Health and Social Care Act for England 2012: the extension of "new professionalism"', *Critical Social Policy*, 33(3): 564–74.

Timmins, R. (1995) *The five giants: a biography of the welfare state*. Harper Collins, London.

Turner, D., Salway, S., Mir, G., Ellison, G., Skinner, J., Carter L. and Bostan, B. (2013) 'Prospects for progress on health inequalities in England in the post-primary care trust era: professional views on challenges, risks and opportunities', *BMC Public Health*, 13(274), at: www.biomedcentral.com/1471-2485/13/274 (accessed 25 July 2015).

Virdee, S. and Cole, M. (2000) '"Race", racism and resistance', in Cole, M. (ed.) *Education, equality and human rights*. Routledge, London.

Webster, C. (1991) *Aneurin Bevan on the National Health Service*. University of Oxford Welcome Unit for the History of Medicine, Oxford.

West, K. (2013) 'The grip of personalization in adult social care: between managerial domination and fantasy', *Critical Social Policy*, 33(4): 638–57.

Williams, F. (1989) *Social policy: a critical introduction*. Polity Press, Cambridge.

12 Whose feelings count?

Performance politics, emotion and government immigration control

Kirsten Forkert, Emma Jackson and Hannah Jones

Introduction

In the summer of 2013, a van with the slogan 'In the UK illegally? Go Home or Face Arrest' was driven through six ethnically diverse London boroughs. The van was part of a pilot scheme, named 'Operation Vaken', carried out by the UK Home Office, the government department responsible for immigration. The stated aim of Operation Vaken was to encourage people whose immigration papers were not in order to come forward and take part in the government's long-standing 'voluntary returns' scheme: to 'go home' to the countries from which they had migrated, with the threat that if they did not, they would be sought out, arrested and deported anyway. This accompanied targeted and high-profile immigration raids on public transport hubs and workplaces, which were publicised on the Home Office's Twitter account with pictures of people being arrested by immigration officers and pushed into the back of enforcement vans, with the hashtag #immigrationoffender. Later that summer, further posters were displayed inside immigration reporting centres in Glasgow and Hounslow, where asylum seekers were faced with images of aeroplanes with the slogan 'This plane can take you home. We can book the tickets.'

This campaign is emblematic of how successive UK governments have chosen strategies that move away from policy discussions about the basis of immigration policy towards demonstrating a 'tough' stance, which is intended to reassure fearful voters that a threat is being contained. However, the tactics used in the 2013 campaign were particularly controversial, even drawing criticism from the UK Independence Party's Nigel Farage and the anti-immigration think tank Migration Watch (BBC, 2013; personal communication). This was in part because of the politically and emotionally charged slogan 'Go Home or Face Arrest'. As Pukkah Punjabi wrote in her comment piece in *The Guardian*, the 'Go Home' slogan evoked National Front graffiti and playground taunts of the 1970s, directed at non-white people whatever their immigration status (Pukkah Punjabi, 2013). The campaign thus raised questions both about who belongs to the body politic but also whose feelings count, reassuring those who are concerned about immigration and provoking fear amongst those who are perceived as outsiders.

This chapter is based on findings from an 18-month research project that investigated the effects of UK government publicity about immigration.[1] The project included working in partnership with activist and charity groups to conduct focus groups with migrants, asylum seekers and British citizens in six areas within the UK (Barking and Dagenham in east London; Ealing and Hounslow in west London; Birmingham and Coventry in the West Midlands of England; Bradford in the north of England; Cardiff in Wales; and Glasgow in Scotland), a national poll, online research, and interviews with policymakers and key local activists in the six areas. Some of these case studies were chosen because they had been specifically targeted by Operation Vaken, which was the case in Barking and Dagenham and Ealing and Hounslow, as well as Glasgow. Others (Birmingham, Coventry and Bradford) were chosen because of their character as super-diverse cities, in which debates on immigration and integration were important locally. Studying Cardiff, as well as Glasgow, also allowed us to consider the role of attitudes towards migrants in the context of devolved government administrations (Scotland and Wales), including the role played by immigration policy in the 2014 Scottish independence campaign. Rather than focusing on Operation Vaken's role in UK immigration control in general, this chapter explores a set of questions about the circulation of emotion in and beyond this campaign. How did the campaign seek to manage the feelings of 'the general public' about immigration control? What were the (unintended?) consequences of this emotion work? How did the campaign fuel (emotional) activist responses? Our analysis therefore examines the assumption that public emotions should or could be managed through such government interventions by also attending to the flipside of this emotion work – the creation of fear, anger and resentment.

Whose feelings? The different values placed on emotions and lives by Operation Vaken

Political discourse at the turn of the century, particularly in the UK, had a particular drive to focus on technical fixes, quantifiable targets and 'evidence based policy' with an emphasis on results that can be measured statistically, claiming a space away from irrational, emotional politics (Levitas, 1998; Fairclough, 2000). Our interest in this chapter is in a government project that overtly set out to work on emotions, claiming to use advertising techniques to fix a tangible problem, as Operation Vaken's stated aim was 'to test whether different communications could encourage any increases in voluntary departures' (Harper and Holbeach, 2013), by circulating messages about voluntary returns in public space – as well as in the less public spaces of migrant reporting centres, places of worship and other venues to which destitute migrants go for support. According to the Home Office's evaluation report, Operation Vaken was intended:

> to test the hypothesis that people without leave to remain in the UK would depart voluntarily if they were made aware that:
>
> •   there was a near and present danger of their being arrested;
> •   the voluntary departure route was explained as an option; and

- safe routes of approaching the Home Office for assistance were provided, without the fear of arrest.

(Home Office, 2013: 2)

So, explicitly, the advertising was intended to provoke fear (of the danger of being arrested), but it was also apparently designed to reassure that there were 'safe routes' through which people without leave to remain in the UK could depart the country without fear of arrest. Though the emphasis in this explanation is on reassurance, the need for reassurance depends on provoking fear. Of course, among people with insecure legal status to live in the UK, a poster and leafleting campaign may not be required to engender fear of being arrested and deported. But the imagery was clearly designed to enhance this fear; the van image in particular included an anonymous border force guard holding handcuffs, and a claim that there had been '106 arrests last week in your area'. The Home Office's explicit statement also ignores the fear that the injunction to 'Go Home' may create for the same people, either because they have fled 'home' or because their life and home are now in the UK. The two choices presented – 'Go Home or Face Arrest' – may equally provoke fear which the promise of a 'safe route' to 'voluntary departure' does not assuage.

Our discussion in this chapter focuses particularly on the very public parts of the government campaign, particularly the 'Go Home' billboard, to consider how the promotion of these messages in public space produce emotional effects which are differentiated, largely depending on the viewer's biography and social position. It is worth noting that the van was only in circulation for a few days and in a small number of boroughs in London. However, its presence became public not just in the physical encounter with the van, which relatively few people experienced, but in its circulation through extensive media coverage, all of which reproduced the van's image and message for a wider public. We argue, drawing particularly on the work of Bridget Anderson and of Sara Ahmed, that these different emotional responses are a technology of governance which both creates and extends political and social divisions. In particular, the attention paid to different emotional reactions of different populations reinforces the boundaries between 'us' and 'them', or between those who are considered part of the general public or 'community of value' and those who are not (Anderson, 2013).

But if Operation Vaken was clearly aimed at provoking fear and consequent actions, it was only openly claimed to have these intentions towards people without leave to remain in the UK. As we will discuss in the rest of the chapter, the emotional effects of the campaign were felt much more widely than among the audience the Home Office claimed to be targeting. The conduct of Operation Vaken appeared to value the emotions of some people – and the material and political consequences of those emotions – over others.

Anderson (2013) provides a detailed analysis of the ways in which the slippery concepts of race, ethnicity and nationalism are intertwined historically, both reinforcing and negating one another. She uses the term 'community of value' to describe the ways in which some people – 'respectable, hard-working families', for example, in the vocabulary of politicians – are seen as constituting the worthwhile part of a national imaginary, as opposed to 'shirkers' or 'scroungers'. This status of value also becomes applied to migrants in differentiated ways, with immigration policy and newspaper headlines equally distinguishing between 'good migrants' – the Australian nurse in the UK, for example – and 'bad migrants' – such as the 'health tourist' using NHS resources to care for their critical illnesses. Those included in the 'community of value' – the 'us' in 'us and them' – are not just of value to the nation, as productive and respectable individuals. They are also the figures whose lives are treated, in policy and in public discourse, as of the most value, as worthy of consideration and care.

Reading directly from the stated aim of Operation Vaken – to encourage people with irregular migration status to leave the UK – the campaign is a clear demarcation of a community of value, and who lies outside of this: people 'in the UK illegally'. However, we argue that the circulation of these messages also draws a less straightforward line of exclusion and inclusion as to who – and whose feelings – are to be considered 'of value'. Sara Ahmed describes the emotional effects of some signs, words and objects as 'sticky'; that is, significance, refer-ences, resonances and feelings can stick to a phrase (such as 'Go Home') despite a change in context or intended or implied meaning (Ahmed, 2004: 91–2). The recognition of a phrase such as 'Go Home' as a racist taunt, one which unsettles the idea of home and reignites a hierarchy of belonging (Back et al., 2012), calls into question for some people – who have been the target or bear the cultural memory of that taunt – their own right to belong in the place they call home. Such stickiness, such resonances and such emotional reactions were not acknowledged by Home Office ministers or others who supported the campaign, as they refused to recognise the symbolic violence of this intervention (Goodhart, 2013; Harper, 2013). The rejection of the emotional reactions (and the politics of these emo-tions) demonstrates, in our view, an exclusion of racialised minorities from the community of value; that is, these reactions are deemed unworthy of consideration.

As we will outline below, reactions to the 'Go Home' van were not all the same. Indeed, many public commentators and research participants suggested that it was intended expressly to appeal to people worried about immigration – to demonstrate that the government was being tough, even, in some sense, to demonstrate to those who felt excluded from the national community of value by economic exclusion that there exists another group that is still further outside of consideration – a yet more abject 'national abject' (Tyler, 2013). This intervention formed part of explicit government drives to make Britain a 'hostile environment' for immigrants (Aitkenhead, 2013), in a context in which ongoing debates about the 'failure of multiculturalism' are taking on new forms that both separate and conflate the racialised hierarchies of belonging in Britain (Back et al., 2012).

Managing fear, performing 'toughness'

Several commentators at the time of Operation Vaken suggested that an important motivation seemed to be to reassure people concerned about immigration control that 'something was being done' by the government (e.g. Behr, 2013; Merrick, 2013; Sdrigotti, 2013). Our research with policymakers confirms this suggestion, with civil servants and policy advisers telling us that while the 'Go Home' billboard was the most extreme artefact to date, the Home Office under Labour as well as Coalition governments had been pursuing the strategy of being seen to be 'tough on immigration' for much longer. While in the early years of Tony Blair's government there was a reliance on providing accurate data to allay fears of immigration (see, for example, Glover et al., 2000), a shift seemed to occur around 2006 and 2007 during John Reid's time as Labour Home Secretary. This seems to be the point at which the received wisdom now common around Whitehall took hold – that 'the public' cannot be persuaded by statistics on immigration as they simply disbelieve them (Duffy and Frere-Smith, 2014). As a result, the emphasis came to be on demonstrating 'toughness' on immigration whether or not 'immigration' was an empirical 'problem'. As one senior think-tank adviser told us:

> they're [the public] not going to believe any immigration statistics. So while the Treasury might be believed on its growth figures, it will never be believed on its economic impacts of immigration. There's more mistrust about immigration . . . With immigration you have every reason to disbelieve data, because the government has told you it's crap at collecting it.
>
> (Policymaker interview, London, June 2014)

So data is not only disbelieved, but this disbelief is compounded by successive governments' communication policies. And attempts to assuage concerns about immigration – which have been stoked both cynically by politicians in electoral campaigns and by governments undermining trust in their own records (Castles, 2010; Gamlen, 2013) – can be seen in stunts such as Home Secretary John Reid accompanying border officials on dawn raids while wearing a flak jacket in 2007 (echoed in visits by Prime Minister David Cameron to view raids on 'beds in sheds' and home raids in 2014); John Reid's rebranding of border control at ports to include reassuring signage and uniforms; current signage in NHS hospitals stating 'NHS hospital treatment is not free for everyone'; and so on.

As in the case of the 'Go Home' van, it is claimed that these efforts aim to show irregular migrants that they will be found (engendering fear), but they are equally instrumental in communicating with people who fear or hate (illegal) immigration. They demonstrate that the government considers (illegal) immigration a problem, and that it is taking measures to deal with it. The word 'illegal' is in brackets in these sentences because in people's understandings of migration in the UK, and in government communications and public debate, the definition of 'migrant' and of the legality of migration is blurry (Anderson and Blinder, 2014). It is most publicly acceptable to be against 'illegal' immigration, as this is seen as a breach of the law (though in some cases legal and illegal status blur as administrative processes and decisions change them). But 'migration' as a whole is something the Coalition government (2010–15) promised to cap, and it is often unclear in their statements on the subject whether they are talking about 'illegal' immigration or immigration in general (see, for example, Harper, 2013). Others have argued that British immigration policies are 'designed to fail', in the sense that governments recognise a need for migrant labour, and an inevitability of immigration, and therefore put in place impracticable immigration controls which can be shown off to anti-immigration advocates, while blaming their failure on wily migrants or outside forces of globalisation (Castles, 2010; Gamlen, 2013).

This opens up broader questions of who 'the public' is that the government thinks it is addressing. As Bridget Anderson (2013) demonstrates, the concern of current governments seems to be with the emotions of those who are considered part of the body politic – people who are seen to be concerned about immigration and in need of reassurance. The classed, raced and gendered nature of these attempts at emotional influence, and of their consequences, is important. In this chapter we focus mainly on the racialised politics of emotion evident in the campaign and responses to it, though elsewhere we continue to explore the interactions of this with class and gender dynamics (see http://mappingimmigration controversy.com for further details).

The following section will explore how Operation Vaken provoked a range of responses within a range of people: migrants with irregular status, those with regular status, racially minoritised British citizens, as well as those who were concerned about immigration. The most common responses, according to our research, were fear, anger and disgust. Some of these fitted with the aims of the campaign (making irregular migrants feel unwelcome) and others were less predictable, reflecting the campaign's resonance amongst other groups.

Emotional reactions and deflections

With Operation Vaken, the Home Office intervened in a cycle of 'emotionality' (Ahmed, 2004) that had both predictable and unpredictable outcomes. This raises questions about whose experiences and emotions count and who is imagined as part of the body politic. As already suggested, the language of 'Go Home' resonated with those who had experienced racism and the taunt of 'Go Home' in the past. The capacity of these campaigns to cause pain through using such language was raised in a debate in the Scottish Parliament, with Labour MSP Hanzala Malik speaking in an emotional register as he described feeling 'shamed, disappointed and shocked' by the campaign. Notably, he said that he did not need evidence that this campaign hurt people's feelings as he had 'lived this evidence'. He recalled: 'I remember when I was young, people used to say "why don't you go home?" but home was Glasgow. They would ask "No, where are you actually from?"' (Scottish Parliament, 2013). This claiming of personal hurt and drawing on past experiences of racism was a powerful moment in Parliament. However, such feelings are denied by Home Office apologists such as journalist and think-tank commentator David Goodhart (2013), who argued 'there is nothing inhuman or racist about encouraging them to come out with their hands up', seemingly unable to recognise these resonances with forms of racism.

For those who were targeted by the Home Office campaign, who had irregular migration status, or who had experienced detention or immigration raids in the past, the campaign instilled fear by prompting recollections of traumatic events. Even for those whose migration status had been regularised, the spectacle of government campaigns still had the power to elicit fear and upset. For example, a focus group participant from Ealing and Hounslow described witnessing an immigration raid at a train station:

> I saw so many UKBA (UK Border Agency) people they were there. I saw them with large dogs, blocking the entire area. I had a visa and have it now also. But I got really scared because I could see the place blocked ... I got so panicked and scared that I went and sat in the wrong train ... When I got on the train I started crying. I was thinking how long will I live with this fear ... I started to think to myself, if I can't move around at all, that people are blocking the way like this, and I'm so scared then perhaps suicide is better.
>
> (Focus group participant, Ealing and Hounslow, August 2014)

This participant was not 'in the country illegally', yet the immigration raid prompts feelings of terror and conjures a sense of being an outsider in a situation that is ongoing and unliveable ('how long will I live with this fear'). This fear then shapes her movement as the public realm becomes too frightening to occupy. As Ahmed argues:

> Fear works to align bodily and social space: it works to enable some bodies to inhabit and move in public space through restricting the mobility of other bodies to spaces that are enclosed or contained ... It is the regulation of bodies in space through the uneven distribution of fear which allows space to be territories.
>
> (Ahmed, 2004: 70)

Thus the emotional reactions of those who feel threatened by these government interventions shape their access to public space, alongside feelings of unbelonging.

These feelings are further complicated by the local context of these public spaces. For example, for participants who were asylum seekers and refugees in cities with long histories of activism around asylum issues (such as Glasgow and Coventry) these moments of fear in public space still occurred but were expressed alongside feelings of belonging to the city and feeling supported by these networks. In the Glasgow case, it is not only the civic but also the national context that impacts on how such anti-immigration initiatives are interpreted. The refugee and asylum seeker participants interpret these campaigns as interventions from a remote and un-Scottish Home Office. In this case, experiences of fear in local spaces are attributed to forces from outside. Thus the geographic context, in particular local histories of migration and activism, impacts on how these interventions in space are experienced and interpreted.

Furthermore, the campaign did not seem to overwhelmingly reassure the general public. In our nationwide survey,[2] we found that, among those aware of the 'Go Home' vans, a higher proportion was concerned that some people were being treated with unnecessary suspicion in everyday situations (34%) than was reassured that the government was taking action against irregular/illegal immigration (28%). In our qualitative research, we found that those who were concerned about the campaign as unfair and racist described feelings of disgust (see below). And even those who expressed strong anti-immigration feelings did not trust the van campaign's effectiveness:

> They're trying to give the impression that they're doing something about it: "We are doing our job, we are catching these illegals, we are putting them in the van and we're taking them to the jail" and half an hour later they're going to let them go again, they're not saying that bit, are they?
>
> (Focus group participant, Barking and Dagenham, August 2014)

Indeed, in our national survey we found that 15% of adults who were aware of the 'Go Home' van said it made them concerned that irregular/illegal immigration might be more widespread than they had realised – that is, the fact that the government was using this intervention actually created fear of uncontrolled immigration rather than reassuring people it was under control.

There are other unintended consequences of such emotional interventions. One reaction among those cast as 'national abjects' (Tyler, 2013) through such campaigns was an attempt to align themselves with 'respectable' citizens. This was done through processes of deflection and boundary-drawing. For example, in a focus group conducted with refugees and asylum seekers in Glasgow, some participants made distinctions between themselves and other groups, including alcoholics, drug addicts, Eastern Europeans and Roma, in order to argue that they were deserving and respectable ('We are normal people, we are not alcoholics or drug addicts'). This was also the case in a focus group conducted in Ealing and Hounslow where focus group participants drew boundaries between themselves as 'respectable' migrants and newer migrants ('There are people that get visas that make their lives here and there are those that come here and get caught up in drink and drugs and they should not be here.') While in the Glasgow case these distinctions were resisted by other members of the group ('actually, I can't say I am better than them'), they serve as examples of how the emotion that is put into motion by the government campaign (against the background of media discourses) can be resisted through deflection, which can in turn add to the stigmatisation of another group.

While the government campaign draws the lines of national belonging of insider/outsider, citizen/non-citizen, our research reveals a more complex set of belongings which impact on how government campaigns are interpreted or viscerally experienced. This can involve harmful words resonating with racist taunts from the past ('Go Home'), the re-living of terrifying experiences ('I'm so scared, perhaps suicide is better') and the deflection of negative emotion onto another group in order to claim national and local belonging. By dividing the population into a 'respectable' general public who can be reassured through such campaigns and alien others who need to be rooted out and expelled, the government discounts a range of experiences and histories of immigration and racism which impact on how people interpret and experience such campaigns. The claim that only those who have something to hide have something to fear ignores these different emotional registers of belonging and experiences of fear.

Our research suggests that when such emotional campaigns are mobilised by the state, the circulation of these emotions outlives the campaigns themselves. For example, our online ethnography showed that the #racistvan hashtag on Twitter (used originally in the angry reactions to the van, as well as in the many online spoofs) remains in use long after the summer of 2013, as the van continues to be evoked as a symbol of overt anti-immigrant sentiment and crude 'dog whistle politics'. In the next section, we will discuss the angry reactions the van provoked in our focus group participants.

Anger and anti-racist politics

Another, perhaps unintended, emotional response to the campaign was that of anger and disgust towards the campaign, which was evident in the reactions from our focus group participants when shown images of the van. These responses were not only reactive, but also involved a sense of dismay and incomprehension at the inflammatory messaging of 'Go Home or Face Arrest', and perceptions that it was wrong for governments to engage in such divisive politics. This was evident in the following response from an asylum seeker focus group participant in Birmingham:

> This makes me sick and it makes me feel as if I want to face these people and know what is their mind behind . . . You cannot just tell people in this country to say you go home.
>
> (Focus group participant, Birmingham, August 2014)

Other focus group participants saw the imagery in the van as explicitly racist (despite claims by the Home Office and other commentators that this was not the case, because it was only targeting irregular migrants) and as inflaming community tensions:

> Personally, that van, I think it never served any purpose whatsoever other than flaunting racism and hatred.
>
> (Focus group participant, Birmingham, May 2014)

The van was also interpreted as exploiting racism and divisive politics for electoral gain:

> What the government is actually doing is using racism, yeah, to fan . . . well they're basically facilitating and fanning racism to get back into power.
>
> (Focus group participant, Coventry, July 2014)

Another focus group participant evoked the cultural memory of a time when both explicit race talk and 'biological' forms of racism which define nationality in terms of ethnicity and birthplace (regardless of one's actual immigration status) were more socially acceptable:

> It makes me feel like no matter what happens, what my outcome is, I will never fit in and become a British citizen, yeah, because of this, "Go home!" because it's reminiscent of back in the days when they used to be blatantly racist towards people.
>
> (Focus group participant, Birmingham, May 2014)

Racism was also evoked (although less explicitly) through questions about why it was acceptable for British people to come to their countries and set up businesses (or, in relation to the history of colonialism, to have taken their resources) but then make the citizens of former colonies feel so unwelcome:

> I'm from India, they ruled my country for 250 years, and they have taken everything. I wouldn't say that all our population are coming here, people who are in need are coming here, then why don't they give the hospitality for the people?
>
> (Focus group participant, Birmingham, August 2014)

What is significant about these angry responses is their identification of what is wrong and unjust (racism, xenophobia, dog whistle politics) – but also that they go beyond being reactive and begin to outline principles of social justice, and to judge the Home Office according to these principles. In *The Cultural Politics of Emotion*, Sara Ahmed draws on Audre Lorde's theorisation of anger and its usefulness for feminism: 'For Audre Lorde, anger involves the "naming" of various practices and experiences as racism, but it also involves imagining a different kind of world in its very "energy"' (Ahmed, 2004: 175). She also argues that 'being against is also being for something but something that has yet to be articulated or is not yet' (ibid.). This is what makes anger a 'subversive counter-emotion' (Flam, 2007: 20), and why it is seen by scholars of social movements as 'the very currency of protest' (Flam, 2007: 26).

Even if they do not fully articulate it, these condemnations of racism and divisive politics also gesture towards a conception of migration based on principles of global social justice rather than 'just keeping the wrong people out' (Walters, 2004: 247). William Walters outlines this conception of migration as one in which countries in the global North take responsibility for the role that the 'regimes of agriculture, trade and finance they have encouraged to their arms industries and security games – play in producing and reproducing the conditions of poverty, distress and conflict which set in motion the movements which [immigration controls] seek to manage' (ibid.). These are not particularly new demands, but they are radical in the context of the current anti-immigration climate of public debate.

In voicing expectations and principles about how governments should behave, these focus group participants – who were a combination of irregular migrants and migrants with legal status – were acting as political subjects, and insisted that their feelings count too, rather than only the feelings of those who are concerned about immigration. This is important, because as migrants they are meant to be objects of state control, but cannot look to the state for any social protection; they are 'at once contained and dispossessed by the state' (Butler and Spivak, 2007: 5). The situation is much worse for asylum seekers and irregular migrants than for those with legal status; however, recent legislation such as the Immigration Act 2014 has also reduced legal rights and entitlements to the welfare state for migrants with regular status (UK Government, 2014). Such developments discourage migrants from feeling they belong to British society or seeing themselves as citizens. Instead, their relationship to the state increasingly becomes only one of enforcement and – on the level of emotions – fear. As discussed earlier, the inflammatory language of the Home Office campaigns evokes experiences of racism for racially minoritised British citizens, suggesting that their feelings are not seen to matter either, or that they matter less than those who are concerned about immigration. For migrants, articulating these sorts of expectations of governments is thus about challenging 'understandings of who can speak, who can occupy political space' (Walters, 2004: 256). It is also about challenging whose feelings are taken seriously.

Conclusion

We have explored the power relations inherent in emotional campaigns such as the one involving the 'Go Home' van. Such campaigns are symptomatic of a climate where some sectors of the public do not trust the government's claims that they are successfully managing immigration, or appeals to rationality such as immigration statistics. In order to restore the perceived lack of credibility, the government resorts to emotional means in order to reassure members of the public who are concerned about immigration and create a sense of unease amongst migrants with irregular status. The 'Go Home' van can be interpreted as an example of this, as an attempt to visibly demonstrate 'toughness' through using the imagery and rhetoric of enforcement, and through high-profile interventions in public space. As a visible demonstration of 'toughness' on immigration, the 'Go Home' van can be interpreted as a response to a cynical political climate in which the public does not trust statistics on immigration, and can be understood as symptomatic of the tensions between the rational and irrational in policymaking. However, the emotional impacts of this attempt by the UK government to manage feeling spread beyond the objectives outlined by the Home Office. We have argued that the campaign as an emotional policy intervention provoked many unpredicted responses, beyond the stated target of making irregular migrants feel unwelcome.

The 'Go Home' campaign can also be understood as both productive of and symptomatic of a political climate in which certain people are considered to belong to the body politic or community of value more than others, and their feelings are seen to matter more than those of others. The theories of Sara Ahmed, Bridget Anderson and Imogen Tyler have been useful in understanding the emotional underpinnings of these sorts of divisive politics which frame some people as belonging to British society more than others, beyond the simple question of whether or not one has legal immigration status.

Responses of fear and hurt demonstrate the importance of interpreting such interventions historically. The language of 'Go Home' echoed experiences and racist taunts of the past, and older conceptions of British citizenship which conflate nationality and race, which are largely no longer socially acceptable but nonetheless still present today. Fear was not only experienced by those who were explicitly targeted by the campaign but also resonated with other groups, some of whom felt they were being told they had no right to be in the country and no right to occupy public space. The campaign fixed a group as 'other' (those 'in the country illegally') and abject, but we found that this led those framed in this way to differentiate between themselves and other 'others' (alcoholics, sex workers, new migrants) in order to claim respectability and the right to belong. Furthermore, emotional reactions were also bound up with activism and resistance. The angry responses were not only reactive but also about specifically questioning such divisive tactics and articulating anti-racist politics. This is one example of how the logic of such governance regimes can be subverted through unexpected emotions (Pykett et al., Chapter 1 in this volume). It indicates how emotions, when used by and against government campaigns, can become a site of struggle.

Notes

1 This chapter is based on research conducted by a collaborative team which included, along with the authors, the following co-investigators: Gargi Bhattacharyya, William Davies, Sukhwant Dhaliwal, Yasmin Gunaratnam and Roiyah Saltus. We would like to acknowledge the use of fieldwork conducted by these colleagues in this chapter, as well as the collaborative input of the whole team into the development of the analysis produced here. We would also like to acknowledge the important contributions to the project throughout its various stages of all of the community partner organisations we worked with (a full list can be found at http://mappingimmigrationcontroversy.com/about/partners). This research was funded by ESRC grant number ES/L008971/1.

2 A survey among a nationally representative sample of 2,424 adults aged 15+ in Great Britain conducted for the Mapping Immigration Controversy Project by Ipsos MORI. Interviews were conducted between 15 August and 9 September 2014. All data is weighted to the known national profile of adults aged 15+ in Great Britain. Respondents were asked which Home Office communications and actions regarding immigration they were aware of and 26% (603 respondents) claimed to be aware of advertising vans around London in 2013 stating 'In the UK illegally? Go Home or Face Arrest.'

References

All website URLs were accessed on 23 June 2016.

Ahmed, S. (2004) *The cultural politics of emotion.* Edinburgh University Press, Edinburgh.

Aitkenhead, D. (2013) 'Sarah Teather: "I'm angry there are no alternative voices on immigration"', *The Guardian*, 12 July 2013, at: www.theguardian.com/theguardian/2013/jul/12/sarah-teather-angry-voices-immigration.

Anderson, B. (2013) *Us and them? The dangerous politics of immigration control.* Oxford University Press, Oxford.

Anderson, B. and Blinder, S. (2014) *Who counts as a migrant? Definitions and their consequences.* Migration Observatory, Oxford.

Back, L., Sinha, S. and Bryan, C. (2012) 'New hierarchies of belonging', *European Journal of Cultural Studies*, 15(2): 139–54.

BBC (2013) 'Farage attacks "nasty" immigration posters", BBC News, 25 July 2013, at: www.bbc.co.uk/news/uk-politics-23450438.

Behr, R. (2013) 'What is it about those vans?', *The Staggers: The New Statesman's rolling politics blog*, 30 July 2013, at: www.newstatesman.com/uk-politics/2013/07/what-it-about-those-vans.

Butler, J. and Spivak, G. (2007) *Who sings the nation state? Language, politics, belonging.* University of Chicago Press, Chicago, IL.

Castles, S. (2010) 'Why migration policies fail', *Ethnic and Racial Studies*, 27(2): 205–27.

Duffy, B. and Frere-Smith, T. (2014) *Perceptions and reality: public attitudes on immigration.* Ipsos MORI, London.

Fairclough, N. (2000) *New Labour, new language?* Routledge, London.

Flam, H. (2007) 'Emotions map: a research agenda', in Flam, H. and King, D. (eds) *Emotions and social movements.* Routledge, London.

Gamlen, A. (2013) 'Tory immigration policy is not doomed to fail – it is designed to do so', *The Guardian*, 14 January 2013, at: www.theguardian.com/commentisfree/2013/jan/14/immigration-policy-designed-to-fail.

Glover, S., Gott, C., Loizillon, A., Portes, J., Price, R., Spencer, S., Srinivasan, V. and Willis, C. (2000) *Migration: an economic and social analysis.* Home Office RDS Occasional Paper 67. Home Office, London.

Goodhart, D. (2013) 'At last we are talking about and dealing with illegals', *London Evening Standard*, 30 July 2013, at: www.standard.co.uk/comment/comment/david-goodhart-at-last-we-are-talking-about-and-dealing-with-illegals-8738189.html.

Harper, M. (2013) 'Racism? It is not racist to ask people who are here illegally to leave Britain', *Daily Mail*, 30 July 2013, at: www.dailymail.co.uk/news/article-2381051/MARK-HARPER-Racism-It-racist-ask-people-illegally-leave-Britain.html.

Harper, M. and Holbeach, Lord Taylor (2013) *Immigration enforcement: Operation Vaken* (written statement to Parliament), Gov.Uk, 31 October 2013, at: www.gov.uk/government/speeches/immigration-enforcement-operation-vaken.

Home Office (2013) *Operation Vaken: evaluation report.* Home Office, London.

Levitas, R. (1998) *The inclusive society? Social exclusion and New Labour.* Macmillan, London.

Merrick, J. (2013) 'Nick Clegg not involved in the "go home" campaign: how the "racist van" is a way to win votes', *The Independent*, 30 July 2013, at: www.independent.co.uk/voices/comment/nick-clegg-not-involved-in-the-the-go-home-campaign-how-the-racist-van-is-a-way-to-win-votes-8738510.html.

Punjabi, P. (2013) 'The day I asked the Home Office to help me go home – to Willesden Green', *The Guardian*, 28 July 2013, at: www.theguardian.com/commentisfree/2013/jul/28/willesden-green-twitter-wind-up-immigrants.

Scottish Parliament (2013) 'Meeting of the Parliament 19th December 2013', at: www.scottish.parliament.uk/parliamentarybusiness/28862.aspx?r=8720&mode=html#iob_78873.

Sdrigotti, F. (2013) 'Migrant bashing as a PR stunt', *Open Democracy*, 12 August 2013, at: www.opendemocracy.net/ourkingdom/fernando-sdrigotti/migrant-bashing-as-pr-stunt.

Tyler, I. (2013) *Revolting subjects: social abjection and resistance in neoliberal Britain.* Zed, London.

UK Government (2014) *Immigration Act 2014*, at: www.gov.uk/government/collections/immigration-bill.

Walters, W. (2004) 'Secure borders, safe haven, domopolitics', *Citizenship Studies*, 8(93): 237–60.

13 Governing through civic pride

Pride and policy in local government

Tom Collins

Introduction

The use and manipulation of emotions have always been a feature of how powerful actors and institutions have engaged populations and energised them in political ways. Political elites have always known that the ability to win 'hearts and minds' is as important, if not more important, than sound policy arguments (Thrift, 2008). Writing as a geographer, I am particularly interested in how emotions and emotive ideas and policies are mobilised in particular places, and how emotions become localised, politicised and contested. Here I think the notion of 'civic pride', and its role in local government policy and practice, provide a useful point of entry for examining questions of emotional governance, and more broadly for analysing how emotions play a role in (and become productive for) local civic agendas. Civic pride is a relatively familiar concept within most local government circles, but it is a term which is rarely given to (academic) critical analysis. My central claim is that by reconceptualising civic pride somewhat, in a more emotionally and politically sensitive way, we can begin to explore some of the underlying meanings and intentions behind current local government policy in the UK, and examine how civic pride connects with a range of contemporary urban issues such as citizenship and belonging, urban regeneration, localism and austerity.

The chapter is organised in the following way. I will first address what civic pride is, means and represents and discuss why it is an important issue for local government. I then examine a range of policies, practices and discourses currently being used within local government across a number of cities and local administrations in the UK, and critically explore some of the emotional, social, economic and political intentions behind them and the existing or future impacts of them. The analysis particularly focuses on civic pride's relationship to civic engagement, urban regeneration and localism, and explores the connections and tensions between emotion, place and policy within local government. I argue that civic pride is shaped by and interacts with a range of emotional, geographical and political ideas and agendas, and that we need to think carefully about what kinds of civic pride are socially and economically productive, who represents and mobilises civic pride, and whether it is, or can be, a politically progressive (or more parochial) discourse and value (and who decides this). The chapter ends with a brief reflection on the overall meaning and impact of civic pride as a tool and symbol of emotional governance.

Theorising civic pride – the self and the city

'Civic pride' describes a sense in which a person feels proud to live somewhere and takes pride in the local community he or she belongs to. As a collective term typically applied to towns and cities, civic pride broadly relates to how local citizens define and represent themselves as a community, how local governments govern and promote places, and how people engage with and intervene in their local area (Shapely, 2012). It is commonly associated with things like lord mayors, civic buildings, grand ceremonies and local enterprise. But in a more emotional sense civic pride usually suggests a strong level of attachment or loyalty to a particular place, and with that a strong sense of identity and belonging (Wind-Cowie and Gregory, 2011). The relatively sparse literature on civic pride has tended to focus on how places promote civic pride to enhance their reputation and promote community spirit, and how, in certain instances, civic pride can be manipulated and exploited by local actors for political and economic gain (e.g. Bennett, 2013; Jones, 2013; Shapely, 2012; Wind-Cowie and Gregory, 2011). However, within this literature there has been very little theorisation or critical analysis of the term itself, both in geography and more broadly in the social sciences, and particularly from a more emotional perspective, in terms of what pride is, or means, in this context. This is perhaps surprising given a long-standing interest within geography around questions of citizenship, belonging and identity and an ever-growing body of work (now) on emotional and affective geographies (Thrift, 2008; see Pykett et al., Chapter 1 in this volume). There has also arguably been a broader lack of interface between emotional geographies and research on the politics of local government – something which this book has sought to challenge, as well as a number of other recent works (see, for example, Bennett, 2013; Jones, 2013). Understanding civic pride from a more emotional perspective allows us to examine how and why civic pride represents an *embodied* feeling and value that is productive and powerful in a range of ways. To understand the ways in which civic pride works, individually, collectively and within local government policy, it is important to understand what pride is and what kind of emotional qualities it represents. These emotional qualities shape the political values that underpin civic pride and impact on the ways in which individuals, communities and governments promote and defend their local identity and autonomy.

Pride is a feeling of self-worth and integrity. It usually signifies a positive and invigorated sense of self, associated with feelings of self-esteem, belonging, success and superiority (Tracy et al., 2010; Smith, 1998). It often reflects a person's or community's (heightened) status or reputation and can – in the positive form – help reinforce people's sense of agency or independence. But pride – or rather the display of pride – can also at times be received negatively by others, as though the proud individual or community is arrogant or overbearing (Wind-Cowie and Gregory, 2011). Pride is of course one of the 'seven deadly sins' within Christian theology. But it has also been an important banner-emotion for a long line of indigenous movements, sexual and racial politics (e.g. gay pride, black pride) and other social movements – following what Neu (2000: 108) describes as a historic evolution from 'a theology of sin to a politics of self-assertion'. Pride, whether expressed in the positive or negative, as a 'virtue' or a 'sin', has a close relationship to its opposite – shame (i.e. a lack of self-worth, a lack of integrity). For in order for one's pride to retain its merit, status or virtue, it must be free of shame, and in many instances the desire for pride may be driven by a (parallel) desire to avoid or overcome shame (Fortier, 2005; Probyn, 2005).

This last point is critical for understanding the potential for pride to be used as an instrument of power. For pride can encourage people to behave in morally or socially acceptable or agreeable ways (perhaps in order to avoid shame), but pride can also be the driving force for transforming a person's or a community's social circumstances, and has a particularly 'aspirational' quality to it. As Tara Smith relates, pride has particularly moral and 'motive' properties and can significantly shape how a proud person thinks and acts:

> Pride is the adoption of this policy toward morality itself. A proud person has clearly-identified values and makes high demands of herself . . . She [*sic*] cares about herself in the healthy sense of making the most of her life – by doing the best, in her actions.
>
> (Smith, 1998: 77)

Through this kind of conceptualisation of pride – as a self-motivating and self-enhancing emotional value – we can begin to think about how civic pride operates within and from a similar psychological and moral basis (Wind-Cowie and Gregory, 2011). In other words, we can begin to recognise how civic pride constitutes a kind of a moral (civic) value of engagement and responsibility, a self-motivating and self-enhancing value that encourages people to engage in where they live and aspire to making it better, more equal or enriched somehow.

In Mattson's (1998: 19) view this form of pride can create 'public-minded' citizens – what he calls a 'proprietary' type of pride related to people's sense of ownership and responsibility over where they live. Mattson's argument echoes a more classical conception of citizenship, in which local citizens are to be judged not just by 'pride' per se (or their feelings about a place) but by their active engagement with where they live – as citizens, as members of a political community (Wind-Cowie and Gregory, 2011; Dagger, 1997). In this way civic pride and civic engagement are reciprocal: if citizens feel proud of where they live, and *take pride* in where they live, they are more likely to be public-minded and develop a greater sense of ownership and responsibility; if people develop a sense of ownership and responsibility, their sense of pride for where they live will likely be reinforced and made stronger.

Alongside this more proprietary type of civic pride, we might also define a more 'celebratory' or identity-based notion of civic pride (Armstrong and Hognestad, 2003; Hunt, 2004). This would primarily concern the ways in which citizens come together and forge a common sense of identity and community spirit. Here things like public spaces, major monuments and buildings, or local events and perform-ances, all become important sites of shared meaning, interaction and engagement – sites in or through which people might share, express or contest civic pride. As I go on to show later, local governments and civic leaders are often keen to promote this kind of civic pride through local policy, in order to foster the belief that the city is one large (imagined) community that thinks and acts together (Dagger, 1997). This is an important point because it raises the possibility for thinking about how certain discourses of civic pride might seek to discipline urban populations in particular, ideologically driven ways, or in ways that to seek to assert difference and distance between those within the city and those beyond or outside it (Darling, 2009).

Indeed, while more could be said here about the different possible types of civic pride, this fundamental dialectic between the *personal* and the *political* (which is also, of course, a dialectic between the *emotional* and the *spatial*) is critical for understanding how and why civic pride forms a tool of emotional governance in cities. Civic pride connects the citizen with the city: it connects the feeling of pride with the virtue of pride; it connects a sense of identity and belonging to a sense of duty, responsibility and engagement. But as I have hinted already, civic pride can also be displayed or communicated in such a way as to suggest that a community is arrogant, pompous or undermining of other people and places. Civic pride – in this more negative (self-serving) form – can also help create a false or misleading image of the city – a city which appears strong, confident and united, but which is actually grounded in a range of inequalities and tensions. We must therefore question what civic pride means, who it represents and serves, and how and why it is being produced or mobilised in a particular place at a particular moment in time. This involves examining to what extent local governments seek to harness or exploit civic pride for particular (ideological) reasons.

Below I discuss a range of recent polices, practices and interventions that have promoted or used civic pride in some way, and use this to assess some of the key questions over the nature, meaning and impact of civic pride in the context of local government. Two key questions shape the analysis here. Firstly, are certain types of civic pride promoted and invested in by local governments, while others are disregarded or maligned? And secondly, is civic pride simply a ruse to dis-cipline citizens and divest responsibility (for, say, welfare and service provision) away from local (and national) government, or is civic pride actually about local self-determination and collective empowerment? In the wider context of changing urban economies, social inequalities, austerity, localism and shifting priorities at national government level, we need to evaluate what is at stake when local governments and local populations promote and defend civic pride and – more importantly – how people's quality of life might be enhanced through or made worse by civic pride.

Civic pride in policy: intervening directly and indirectly

Civic pride's role within local government policy can be examined in two broad ways. One is to look at examples in which civic pride has been the explicit aim and aspiration of the policy itself, or where a certain idea, or project, has been promoted under the banner of 'civic pride'. Another is to explore the role of civic pride within other policies, as a kind of supporting or enabling idea or discourse; this includes policies and initiatives introduced by central government but implemented locally. With this kind of conceptual division in my mind, I examine both the direct and indirect ways in which civic pride has been promoted and defended by local governments, as well as how local governments have (in some cases) resisted central government policy (and other government decisions) *out of* a sense of civic pride.

Direct interventions into civic pride

In recent years there has been a wide range of policies, practices and interventions that have explicitly promoted civic pride, or have at least been mobilised under the banner of civic pride across local government. Many are not so much 'new' policies as they are iterations of older forms of local policy-making, remobilised and realigned to fit present-day circumstances. This of course builds upon a long history and heritage within Britain – one thinks particularly of the Victorian period in this regard – of enterprising local councils and boroughs that have sought to promote civic pride and local enterprise (Hall, 1997; Harvey, 1996; Shapely, 2012). Most UK city councils and local authorities currently have, if not a discrete 'civic pride policy' per se, some kind of initiative or ambition around civic pride. Nottingham City Council, for instance, has as its official city slogan 'A Safer, Cleaner, More Ambitious Nottingham: A City We're All Proud Of' (Nottingham City Council, 2015). Durham City Council has a 'Civic Pride Team' with 'Civic Pride Officers' that deliver a range of initiatives around community engagement and neighbourhood renewal (Durham City Council, 2015). The London Borough of Richmond has its own 'Civic Pride Fund', funding a range of community projects across the area. The fund's two main priorities, according to the council's website, are: firstly, to develop Richmond as 'a borough to be proud of – making public places more attractive, enjoyable and distinctive and building stronger communities', and secondly to use 'the skills and talents of local people to benefit the local community' (London Borough of Richmond, 2015).

East Staffordshire Borough Council in the West Midlands meanwhile has recently released its own 'Civic Pride Strategy' for 2014–17. The strategy's vision was framed around the following descriptor:

> East Staffordshire is a great place in which to live and work and to visit – it has great assets that contribute to the well-being of its people. The people of East Staffordshire recognise the value of these assets and embrace a collective responsibility for helping ensure their towns, villages and open spaces are kept clean, green and safe for future generations; that they are places of which we are all proud and that provoke a sense of civic pride.
>
> (East Staffordshire Council, 2014: 10)

The strategy is based around a number of key priorities: improving the physical fabric of local areas (particularly in relation to litter and fly-tipping), raising awareness of local issues, promoting and rewarding participation, and inspiring future generations of the area. Although some of the key aims and aspirations of the strategy are not unique to East Staffordshire (one could conceivably replace 'East Staffordshire' with any place in the quotation above), it is relatively unusual to see a local authority develop such an explicit and wide-reaching local strategy for civic pride. The positive and reinforcing language of the document – that East Staffordshire is a 'great place' that has 'great assets', that it is a place that embraces 'collective responsibility' on behalf of 'future generations' – is clearly written in such a way as to acknowledge and affirm people's pride in the area and to harness this pride for the benefit of the local community. While neither pride nor civic pride is explicitly defined or explained in the document (a not uncommon feature of many local policy documents, I have found), it is clear that the council aims to promote some notion of civic pride (a 'proprietary pride' perhaps) across East Staffordshire, marrying the *feeling* of pride with the *virtue* of civic pride. The council anticipates that such interventions can help create a 'clean and safe environment' (2014: 3–11) and 'stronger communities', and contribute to more 'personal happiness'; they may also – and perhaps most importantly – help save the council significant amounts of money. As the strategy explains, if people show more civic pride and look after their local environment, then the costs associated with, say, street cleaning or fly-tipping will be greatly reduced – because people will in theory be less likely to litter or fly-tip (East Staffordshire Council, 2014: 12). In other words, a 'prouder' community will not only help improve the quality of the local environment and people's well-being, but also reduce unnecessary costs and financial burdens on the council. Whether this means that local governments can therefore – as a result of this civic goodwill – justify cuts in local services or cuts in public-sector jobs (on the basis that people's commitment to civic pride will somehow 'fill in the gaps' and create less dependency upon local municipal services) is another matter and, as I suggest later, an important one in terms of assessing the broader politics of civic pride.

This link between civic pride and economic gain is important not just in terms of local governments trying to save money, but also in terms of trying to make money. There is a significant literature within urban geography that has considered the role of civic pride within urban regeneration and marketing strategies (Harvey, 1989; Hall, 1997; Jayne, 2012). This has tended to be grounded in Marxist ('structural') types of analysis, with fairly limited attention to the role of emotions (except, for example, Bennett, 2013 or Jones, 2013). Urban geographers have brought attention to how civic pride has become an important part of the rhetorical arsenal of local government in a post-industrial ('entrepreneurial') era, where civic pride is used to help sell the city and legitimate investment in urban redevelopment and regeneration projects. Typically, local pride and local identity are promoted in order to help create a positive (forward-looking) image of the city – one which represents confidence, success, 'buzz' and ambition (Bennett, 2013; McCann, 2013).

Geographers have tended to be critical of the ways in which local governments sometimes exploit the language and appeal of civic pride in order to sell and 'sell off' the city to private, mobile capital (Harvey, 1989; Hall, 1997; McCann, 2013). In this reading, civic pride (and all the marketing materials, strategy documents, promotion videos, advertising banners and media campaigns that cities typically use to portray and promote civic pride) can help fabricate an urban image that is void of internal tensions, conflicts and injustices and builds a false narrative that urban regeneration will benefit the city as a whole (Harvey, 1989, 1996; Boyle, 2011). As Boyle observes, this kind of strategic mobilisation of civic pride, as a tool of civic 'persuasion' effectively, can therefore be used to help legitimate neoliberal change and deflect attention away from the uneven and often pernicious effects of speculative urban development:

> place marketing projects come to double as civic boosterist projects; hallmark events serve both to sell places and to mobilise and manipulate local jingoism and civic pride to galvanise support for local accumulation strategies. The mantra of 'community' and 'locality' is deployed to contain, subdue and conceal the inequalities, risks and threats which are generated by local economic development initiatives – a modern-day version of bread and circuses.
>
> (Boyle, 2011: 2674)

While it may be important to point out how civic pride might be manipulated and exploited by local governments in the ways Boyle and others describe (e.g. Harvey, 1989), there is equally a danger here of overstating the well-rehearsed 'neoliberal critique' and dismissing civic pride as a rather facile urban marketing term, void of its potential to create alternative forms of growth and a more engaged citizenry. It is important not to forget of course that local populations produce and contest their own versions of civic pride, their own narratives, values and traditions, which may be rather different from how local government constructs and promotes civic pride. As Jayne (2012) has shown in the city of Stoke-on-Trent, for instance, more 'traditional', 'parochial', 'working-class' notions of civic pride and civic identity can often co-exist with, and rub up against, supposedly more 'progressive', 'entrepreneurial', 'middle-class' notions of civic pride, in ways that can divide local communities and obstruct local government change. Directly intervening in civic pride through policy can therefore represent a significant challenge for local government, since it often requires negotiating across a range of values, priorities and expectations – indeed it has to, in theory, establish the right kind of pride, in the right place, at the right time (Fortier, 2005; Jones, 2013).

Indirect interventions – civic pride and the issue of localism

Alongside these more direct measures, there have also been more indirect ways in which civic pride has been a feature of recent UK local government policy. As

I outlined earlier, I am primarily referring here to examples in which civic pride has been used in support of, or as enabling mechanism for, wider policies and interventions, at both local and national level. While there are a variety of possible examples one could raise here, I want to focus on the new 'localism' agenda formed under the UK Conservative–Liberal Democrat Coalition (2010 to 2015). The concept of localism itself – as a principle of local communities being able to exercise and defend certain rights and freedoms over where they live – is hardly new of course, and one could argue localism represents much of what civic pride has historically stood for (i.e. local power) (Hunt, 2004). The passing of the Localism Act in 2011 nevertheless became a key milestone in the realisation of localism as a (modern) parliamentary ambition and reflected a significant – and now, it seems, growing – shift towards greater devolution and regional autonomy within the UK (Featherstone et al., 2012; North, 2011). On the back of concerns that the UK had become 'over-centralised' in its approach to local public spending and economic growth, the purpose of the Act was effectively to decentralise a range of new powers and responsibilities to local governments (including new tax powers, planning controls and controls over managing community assets), and to empower local communities to manage more local services and take greater control over local infrastructure and neighbourhood planning. Local governments and communities are now better protected in law to make decisions that matter locally, while local democracy has been (at least in theory) strengthened by measures to make it more transparent and fair – for instance, communities now have a new 'right to challenge' over local authority planning decisions, while councillors have also come under new 'predetermination' rules to encourage more open debate over planning and development decisions, protecting them from excessive legal challenges from developers or private companies (DCLG, 2011).

The politics of localism is complex and echoes many of the issues raised above about the politics of civic pride (see Clarke and Cochrane, 2013 for a cogent review). While some commentators have lauded the principles of localism on the basis that it offers local governments and communities greater autonomy and accountability over what decisions are made and how resources are allocated, others have been highly critical of the Coalition's agenda, because it has come at a time of austerity (Westwood, 2011; Featherstone et al., 2012, Lowndes and Pratchett, 2012). Austerity, and the ongoing impacts of the post-2008 recession, have forced UK local authorities to make highly challenging (and politically damaging) decisions around pay cuts, job losses and reduced welfare and local service provision. One might say that civic pride has suffered badly as a result, although for many this may well be immaterial to the deeper impacts austerity has had on living standards and social security. The irony of the new localism agenda has been to remind people and places of their very dependency on government welfare and the national economy – as though localism has become a euphemism for 'sink-or-swim' (Smith Institute, 2014; Newman, 2014).

As Featherstone et al. (2012) and Clarke and Cochrane (2013) have suggested, while localism may sound attractive in principle, and may in theory provide the legal foundations for a more engaged and empowered democracy in the UK, it produces its own political and economic straitjacket for local government. This straitjacket is rooted in the illusory idea that the 'local' is the scale at which structural urban issues can really be resolved, and that, particularly at a time of recession and austerity, local governments have the capacity to deliver (sustainable, equitable) economic growth and address local inequalities. Featherstone et al. (2012), for instance, stress how issues around the economy, the NHS, the welfare state, immigration and multiculturalism are profoundly *inter-scalar* in nature and cannot be resolved simply at the local level. Clarke and Cochrane (2013) meanwhile warn of localism's more divisive and destructive consequences:

> the relentlessly parochial nature of the local is said to invite fragmentation, not only limiting the ambitions of those engaged in politics at that level, but also encouraging division and competition between those who should be united in the face of global challenges.
>
> (Clarke and Cochrane, 2013: 10)

The politics of localism therefore draws parallels with the politics of civic pride, particularly in terms of the boundaries between a more 'parochial' world view and a more 'progressive', outward-reaching world view (and whether these are mutually exclusive or can co-exist). With these kinds of issues in mind, we can briefly consider here how a more 'progressive' sense of localism might connect with a more progressive sense of civic pride, and the kinds of political and ethical questions this raises for local governments.

Janet Newman (2014) has argued that local governments operate across 'landscapes of antagonism', anchoring and channelling a range of competing and contradictory ideological projects, across a range of scales and relations. While they may represent and govern that which, in political and legal terms, is local and civic, these domains themselves depend greatly upon that which lies beyond the local and the civic (national markets, for instance) and therefore beyond local government control. As Doreen Massey has famously argued, what counts as 'local' is in fact the coming together of a range of internal and external relationships that *produce the local* as something which is contingent, relational and necessarily dependent upon the non-local (Massey, 1994; Darling, 2009). While this point warrants a more extended discussion than I can offer here, this suggests that a more 'progressive' sense of localism, like a more progressive sense of civic pride, might recognise that the local and the civic should be conceived in ways that are not antagonistic to the 'outside' (whether that means other people, other places, other institutions), but rather are produced through, dependent upon and in solidarity with the 'outside'. In other words, places, like people, should not become so proud, so self-absorbed, as to isolate themselves or resist co-operation with others; nor should pride make local communities or governments think that their way of doing things is necessarily better than or superior to the way others do things (Wind-Cowie and Gregory, 2011). Thinking through a more progressive kind of localism therefore opens up space for thinking about how local areas might form new types of alliances with other cities and other regions in an attempt to address inequalities, share collective (civic) resources and facilitate new types of pride.

To end this chapter, I want to briefly discuss how the fight against austerity, across local government, demonstrates one possible example of a more 'progressive' kind of localism emerging – albeit a highly antagonistic one (Newman, 2014). We have witnessed a number of local authorities, particularly Labour-run councils and particularly councils in the north of England, banding together in recent years to protest and campaign against regional inequalities and the impacts (and potential dangers) of the cuts (Smith Institute, 2014). In December 2012, for instance, the Labour-led councils of Newcastle, Liverpool and Sheffield wrote a joint letter in *The Observer*, warning that the 'unfairness of the government's cuts is in danger of creating a deeply divided nation' (*Observer*, 2012). Fearful of a repeat of the kind of rioting a number of cities witnessed in 2011 (the so-called 'England Riots'), they urged the government 'to stop what they are doing now and listen to our warnings before the forces of social unrest start to smoulder'. Like many other local government-led invectives since then, this letter was a clear stand against austerity and a clear push by civic leaders to work together to protect local welfare and local infrastructure. Even Conservative councils have, more recently, started to denounce the extent of austerity within local government, as the cuts begin to impact on local infrastructure, social care and local leisure facilities (see *Guardian*, 2014). These calls against austerity (tinged no doubt with feelings of anger and frustration – shame perhaps – over what is happening to the public sector) will perhaps not represent the kind of positive civic pride message that the growth-hungry 'neoliberal city' needs, and nor will they necessarily serve to ignite a more active and engaged citizenry (although we have witnessed a large number of anti-austerity/anti-cuts groups and networks emerging in the past few years). But they do serve as an important mark of credibility and character for local government, of civic leaders and councillors showing that they *take pride* in their cities and local communities, that they are prepared to stand up for local people and are willing to engage with an anti-austerity movement that is reaching beyond the confines of the local to protect the wider civic good.

Conclusion

This chapter has focused on the emotional and political geographies of civic pride, and how local government policies are produced, mobilised and contested in the name of civic pride. Civic pride relates to how local governments and local communities express, promote and defend local identity and autonomy, as well as the different ways people engage with and intervene in local civic life. I have shown how civic pride can be: 'proprietary', in the sense that civic pride reflects or affirms people's sense of ownership and responsibility over where they live; it can be 'celebratory', in the sense of bringing a community together around shared histories, shared traditions or shared sites and spaces, and it can serve as a marketing tool and political vehicle for urban regeneration. Civic pride can also generate different views about its political and ideological character – it can at times appear 'parochial' or 'conservative' and at other times 'resistive' or 'progressive', depending on the context in which it is being expressed or mobilised and who it is representing. It is therefore a multi-faceted, multi-purpose concept or 'tool' that cities and communities collectively embody and use in different ways, reflecting and exposing a number of implicit tensions, fractures and contradictions.

In contributing to the themes of this book, I have shown how civic pride represents a particular from of 'emotional governance' that harnesses the power of emotions to encourage people to act in civic-minded and locally productive ways. Civic pride ties together the emotional and the political in the context of the local and, as I have shown in this chapter, can be mobilised both directly and indirectly in policy to govern, discipline, persuade and empower local citizens for a variety of purposes. Local governments can also use or mobilise policy to increase or strengthen civic pride and (re)engage citizens with the/their city and the civic sphere. Not all policy interventions are or will be successful, however, and nor will they necessarily have their desired effect upon local communities (in some cases they may actually damage or undermine people's sense of civic pride). We should therefore consider the different 'pride-on-pride' dynamics that unfold in places and the different ways people contest and resist *with* pride – whether that occurs locally, between local citizens and local authorities, or regionally and nationally, between local and central government. Future research in this area might consider how different scales of pride (local, regional, national, for instance) fold together, and whether pride at one scale can be productive for, or disruptive to, pride at another scale (see for some discussion: Wind-Cowie and Gregory, 2011). Civic pride is the battle of and for emotions in the city, but its implications reach far beyond the city. This is why civic pride will continue to form an important point of geographical debate, shaping the meaning and significance of where we live, how we live and the extent to which the local and the civic define who we are and what we value.

References

Armstrong, G. and Hognestad, H. (2003) '"We're not from Norway": football and civic pride in Bergen, Norway', *Identities: Global Studies in Culture and Power*, 10(4): 451–75.

Bennett, K. (2013) 'Emotion and place promotion: passionate about a former coalfield', *Emotion, Space and Society*, 8: 1–10.

Boyle, M. (2011) 'The new urban politics thesis: ruminations on MacLeod and Jones' Six Analytical Pathways', *Urban Studies*, 48(12): 2673–85.

Clarke, N. and Cochrane, A. (2013) 'Geographies and politics of localism: the localism of the United Kingdom's Coalition Government', *Political Geography*, 34: 10–23.

Dagger, R. (1997) *Civic virtue: rights, citizenship and republican liberalism*. Oxford University Press, Oxford.

Darling, J. (2009) 'A city of sanctuary: the relational re-imagining of Sheffield's asylum policies', *Transactions of the British Institute of Geographers*, 35(1): 125–40.

DCLG (Department of Communities and Local Government) (2011) *A plain English guide to the Localism Act*. DCLG, London.

Durham City Council (2015) *Civic pride team*, at: www.durham.gov.uk/civicpride (accessed 25 April 2015).

East Staffordshire Council (2014) *Civic pride strategy 2014–2017*. East Staffordshire Borough Council Publications, East Staffordshire.

Featherstone, D., Ince, A., MacKinnon, D., Strauss, K. and Cumbers, A. (2012) 'Progressive localism and the construction of political alternatives', *Transactions of the British Institute of Geographers*, 37(2): 177–82.

Fortier, A.M. (2005) 'Pride, politics and multicultural citizenship', *Race and Ethnic Studies*, 28(3): 559–78.

Guardian (2014) 'Council cuts: local Tories lead criticism as "savings" hit vital services', *The Guardian*, 18 December 2014, at: www.theguardian.com/society/patrick-butler-cuts-blog/2014/dec/18/council-cuts-local-tories-lead-criticism-as-savings-hit-vital-services (accessed 19 May 2015).

Hall, T. (1997) 'Images of industry in the post-industrial city: Raymond Mason and Birmingham', *Cultural Geographies*, 4(1): 46–68.

Harvey, D. (1989) 'From managerialism to entrepreneurialism: the transformation in urban governance in late capitalism', *Geografiska Annaler B*, 71(1): 3–17.

Harvey, D. (1996) *Justice, nature and the geography of difference*. Blackwell, Oxford.

Hunt, T. (2004) *Building Jerusalem: the rise and fall of the Victorian city*. Henry Holt and Company, New York.

Jayne, M. (2012) 'Mayors and urban governance: discursive power, identity and local politics', *Social and Cultural Geography*, 13(1): 29–47.

Jones, H. (2013) *Negotiating cohesion, inequality and change: uncomfortable positions in local government*. Policy Press, Bristol.

London Borough of Richmond (2015) *Civic pride fund*, at: www.richmond.gov.uk/civic_pride_fund (accessed 19 May 2015).

Lowndes, V. and Pratchett, L. (2012) 'Local governance under the Coalition Government: austerity, localism and the Big Society', *Local Government Studies*, 38(1): 21–40.

McCann, E. (2013) 'Policy boosterism, policy mobilities, and the extrospective city', *Urban Geography*, 34(1): 5–29.

Massey, D. (1994) *Space, place and gender*. Polity Press, Cambridge.

Mattson, K. (1998) *Creating a democratic public*. Pennsylvania State University Press, Pennsylvania.

Neu, J. (2000) *The tear is an intellectual thing: the meanings of emotion*. Oxford University Press, Oxford.

Newman, J. (2014) 'Landscapes of antagonism: local governance, neoliberalism and austerity', *Urban Studies*, 51(15): 3290–305.

North, P. (2011) 'Geographies and utopias of Cameron's Big Society', *Social and Cultural Geography*, 12(8): 817–27.

Nottingham City Council (2015) Homepage, at: www.nottinghamcity.gov.uk/article/22779/Homepage (accessed 25 April 2015).

Observer (2012) 'City council leaders say deeper cuts will spark civic unrest', *The Observer*, at: www.theguardian.com/politics/2012/dec/29/cuts-councils-newcastle-liverpool-sheffield (accessed 29 January 2015).

Probyn, E. (2005) *Blush: faces of shame*. University of Minnesota, Minneapolis.

Shapely, P. (2012) 'Civic pride and policy in the post-war city', *Urban History*, 39(2): 310–28.

Smith Institute (2014) *Labour and localism: perspectives on a new English deal*. The Smith Institute, London.

Smith, T. (1998) 'The practice of pride', *Social Philosophy and Policy*, 15: 71–90.

Thrift, N. (2008) *Non-representational theory: space, politics, affect*. Routledge, New York.

Tracy, J., Shariff, A. and Cheng, J. (2010) 'A naturalist's view of pride', *Emotion Review*, 2(2): 163–77.

Westwood, A. (2011) 'Localism, social capital and the "Big Society"', *Local Economy*, 26(8): 690–701.

Wind-Cowie, M. and Gregory, T. (2011) *A place for pride*. Demos, London.

14 An affective journey to active citizenship

Mark Griffiths

Introduction

British Government social policy has since the 1980s sought to promote the idea of active citizenship. By various means – Housing Associations (Kearns, 1992); Neighbourhood Watch (Clarke et al., 2007); the Citizen's Charter (Fyfe, 1995); the Crick Report (Haste, 2004); the National Health Service (Poole, 2000) and education (Ross, 2007) – policy makers, via collaborations with civil society (and, increasingly, private) actors, have worked hard to impress on people the importance of becoming active citizens. Much of the academic research in this area has drawn on Foucauldian ideas of governance to track the 'self-reliance' of citizens as part of the 'neoliberal reconfiguration' of citizenship (Barnett, 2002). In these commentaries, discursive arrangements of the world normalise strategic rationalities to render active citizenship a process of 'self-responsibilisation' practised by rational subjects making (more or less) rational choices (Rose, 1996; Clarke et al., 2007). This chapter seeks to complement this body of literature by exploring a potentially new front in the promotion of active citizenship. The aim is to focus on the non-rational: to reimagine the UK government's Big Society vision of 'a lifelong commitment to active citizenship' (Conservative Party, 2009a) from the point of view of a pre-rational subject whose world unfolds 'below cognition and consciousness and beyond reflectivity and humanness' (Pile, 2010: 8).

The spectre of governments managing populations in spaces below cognition is a disconcerting prospect, and the concepts of 'neuroliberalism' (Isin, 2004), 'affective politics' (Ahmed, 2004) and 'affective governance' (Hook, 2007) have become important tools in the project to understand the creative and pervasive ways that our 'affective capacities and relations are the "object-target" of techniques of governance' (Anderson, 2012: 30). For geographers this has meant developing a 'sharper geographical sensibility to things subterranean' (Anderson and Smith, 2001: 9) in an attempt to track the 'unseen' world that affects us (see Dewsbury et al., 2002). In this affective world, the 'unseen' constitutes extra-linguistic phenomena that, plainly, resist representation; we can speak of emotions more easily than we can of affects. Used evocatively, however, language can carry emotions and affects and therefore becomes an important tool in the discussion and understanding of affective life. Reflecting this, to 'get at' the affective, extra-linguistic world this chapter takes leave of certain writing conventions, presenting a narrativised affective journey to active citizenship. I use the case of International Citizen Service (ICS), a British government volunteering programme that sends young people abroad to work on development projects and requires volunteers to continue their work on return to the UK. The account re-evokes the affective tenor of the ICS programme's marketing and pedagogical material. The material was examined using the body 'as a recording machine itself', keeping faith that something of my own 'recorded' embodied experience can, as John-David Dewsbury has argued, 'become legitimate data for dissemination and analysis' (2010: 327).

By way of conclusion I discuss issues to do with methodology, imagination and the conceptualisation of affect and subjectivity, acknowledging potential criticisms while also arguing that such creative accounts can contribute to the project of better understanding the ways that our affective capacities are targeted by techniques of governance.

An interruption

The scene is a library somewhere in the UK where a 21-year-old student is studying for his final exams. His days have become mundane routine; he's bored and aches for distraction. He stares and fidgets, restlessly, unthinkingly; his body craves movement. His government is about to oblige via a series of 'affective pushes' that will breathe life into his routine. Taken independently they pass with little impression, but considered cumulatively these pushes begin to orientate him on a journey: his body gradually moves towards a government-run volunteering programme that – he does not yet know this – aims to instil in him 'a lifelong commitment to active citizenship' (volunteerics.org). This is a story of a government circulating affects and managing citizens; it is a story that begs an openness to the sensuous, to something always beyond grasp and a willingness to feel a subtle push to the world.

It begins quite innocuously one slow morning in the library where – as always – his concentration is staccato. He listens to Choice FM – 'London's No. 1 urban radio station' (*The Guardian,* 2013) – whose noise marginally edges books in a languid contest for his attentions. The radio seldom emerges from the background until it begins transmission on a non-verbal channel. This arrives in the form of an advert that opens with 'the sounds of walking through rural Africa'.[1] Time slows. The studio-produced crunch of dry twigs underfoot triggers something in him, takes him back, and transports him on 'a long walk to help villages in Tanzania to build a clean water supply'. He's never been to Tanzania but it's there in his felt past; he's part of a post-Band Aid generation that grew up in 'affective intimacy' with the Poor of the South (Somalia, Rwanda, Sudan) (Lousley, 2014). Those images of pot-bellied kids left him physically uneasy. *Kwashiorkor*, he doesn't know the word (who does?) but his body *feels* it, and its pre-cognitive autonomic reactions precede and precipitate – the body is always 'faster than the word' (Massumi, 1995: 86) – his feelings of shock, discomfort, unease. The advert probes at these reactions once again, the feelings follow shortly after. It's a visceral routine that – thanks to Comic Relief, Oxfam, Live8 – has 'accumulate[d] in memory . . . in reflex' (Massumi, 2002: 213) to become part of his 'affectively imbued memory bank' (Connolly, 2002: 71).

His media-formed experiences of poverty are carried somatically and the lack of potable water in Tanzania rouses the affective remnants of those pot bellies and the autonomic responses again put him at unease. The commissioned voice capitalises, assuring – 'you don't need cash or qualifications . . . just the motivation to make a difference' – and luring – 'help fight poverty' – him. If we're to give his emerging state a name, it would be something close to empathy, perhaps even shock, but we can't: we're in a mysterious realm 'of the missing half-second' before affects are nameable (Massumi, 1995: 89). Whatever it is, it weighs heavy, lowers him, until he's presented with an imperative to hope: 'change your world – apply online at volunteer ICS dot org' (RACC, 2013). It's an animation that resonates through each nerve, relieving the weight and completing a potent fusion of lowering and lifting affects; the advertisement is perfectly attuned to his affective sensibilities. Quite suddenly, the extraordinary has permeated his quotidian ordinariness, it carries him away from his books and dares him to imagine distant times and places. The prospect of travelling to *Africa* tingles, it reverberates through his body and lifts him to a state of readiness, ensuring the words – *volunteer I . . . C . . . S . . . dot org* – are etched on his memory. Whosoever initiative (the Conservative Party), whoever the marketing executive (radioexperts. co.uk) and whoever the financier (Department for International Development. DfID), their collaboration amounts to the transmission of affects that disrupt stasis. They course through him, priming him to act (Thrift, 2007: 26).

He's moved enough to click through volunteerics.org. It's a volunteering programme to fight global poverty. He follows some links, opens tabs: various press articles on current issues, something from the DfID YouTube channel, something else from gov.uk. The last one looks dull but some of the articles take his eye. There's one by his Prime Minister writing in *The Guardian*. He reads and is immediately struck: 'in the time it takes to read this article, 15 children in the world's poorest countries will die from a preventable disease like diarrhoea or pneumonia. We would not stand for that at home' (Cameron, 2011). It's a visceral development imaginary; the Prime Minister asks him to consider global inequalities via intensely affective figures: the fear of disease, the hope of prevention and the pride of home. It's well known that these – fear, hope and pride – are each pervasive affects *par excellence* in the business of 'making things happen' (Massumi, 2005; Anderson, 2006; Anderson and Adey, 2011). He clicks on an embedded link that takes him to some government statistics on development:

> 9.2 million children die before the age of five each year. Two million die on the day they are born – and 500,000 women die at childbirth. A third of children in Africa suffer brain damage as a result of malnutrition. 72 million children are missing out on an education. Every day 30,000 children die from easily preventable diseases. That's 21 children every minute. Thirty three

million people are infected with HIV/AIDS. There are 11 million AIDS orphans in Africa. Every hour, 3,000 people become infected with HIV and 225 people die from AIDS . . . and 25 of these are children.

(Conservative Party, 2009b: 8)

'Fucking hell, that's awful.' He swears because eloquent is not how he feels; words fall short. It's not an unusual reaction, rather a sign of an affective presence on the edge of cognition 'that talk cannot grasp' (Katz, 1999: 4). Prompted by his shock, a mix of Penelope Cruz, Brad Pitt and emaciated bodies run through his mind: the 'click advert' from when he was a teenager, the deaths of poor children are spaced at regular intervals. *Click, click, somebody dies*. The similarity with the circulated 'suffering statistics' (Sharpe et al., 2010: 1130) is uncanny, and his body is once more called on to live those affects. Obligingly, he watches on YouTube. The rawness of the imagery and the direct address is intense; he can only be moved: 'a child dies every three seconds of extreme poverty. And you? What are you doing?'[2] 'Nothing,' he mouths guiltily.

He clicks back and reads on to find his government recycling the very same intensities:

How could this have been tolerated? Our grandchildren will ask us: what did you do to help tackle poverty? They will wonder at the absurdity of a world that could put men on the moon, but still let millions of people suffer and die from easily-preventable diseases.

(Conservative Party, 2009b: 8)

Again he's positioned personally in matters of global poverty and, crucially, reminded of his inaction in an address that cannot but compound his rising sense of guilt. And as strong as this was before, it's this time of a different order as he's asked to consider his contribution under the expectant gaze of his not-yet-conceived offspring of offspring. This rhetorical move aims squarely at the pre-cognitive dimension of familial instinct and of course he's wired to respond to the copywriter's carefully crafted image of dying children. Then the literary language is ramped up, drawing him in further:

Every time the candle of light is snuffed out by disease, we all suffer. Every time ignorance triumphs over enlightenment, we are all made injured. Every time a child is born into a cycle of poverty, we are all made poorer.

(Ibid.)

The images arrive in rhetorical threes, each upping the emotional ante. He's figuratively positioned within the text as part of 'we', part of the 'cycle of poverty', and with the affective figure(s) of his future offspring still pulsing, he's pushed to extend from the temporal to the spatial through (yet more) evocations of distant dying children. Animated by these affecting spectres, there's a visceral rawness to the anxiety he begins to feel at *their* suffering and *his* inaction.

It's only been an hour but already the affectively freighted construction of development has roused anxieties and ignited compassion (Sharpe et al., 2010). The here and now of the library quivers with the affective intensities of distant places, peoples and prospects. It seems remote, but it's not: this non-verbal communication of global poverty is for him (and for many of his peers) the first in a series of affective pushes towards active citizenship.

'Make a difference – apply now'

Over the following days discomfort flickers through him as the images of dying babies remain. His inaction is a source of anxiety and it eats away at him, a constant reminder of the possibility of action: sign up, volunteer. It's constantly with him – the promise of 'something more, a more to come' (Massumi, 2002: 215) – buzzing in the background, ready to emerge at the subtlest cue. And the cues are everywhere thanks to targeted advertising such that even the algorithms seem to be in sync with his body. There's a banner on his Facebook newsfeed: 'ICS, make a difference – apply now'. He clicks and enters the site's image-rich sensorium that at every turn invites him to 'fight poverty' as a 'global active citizen' (volunteerics.org). Warmth radiates from the screen in a series of photographs of volunteers and hosts. He's drawn into the images in a co-constitutive entanglement – 'to look is to have one's affective state changed' – that 'send[s] [him] along emotive pathways' (Joffe, 2008: 86). The marketing's corporeal literacy at this point shows itself to be plenty capable of appealing to his evolutionary affective dispositions (see Thrift, 2004: 72–4). The images of the Programme (without exception) consist of volunteers surrounded by (poor, of course) children – and there's a surfeit of *genuine*, smiling faces (he doesn't count but there are thirteen images populated by 26 different faces). He knows their happiness to be genuine because he perceives the *zygomatic major* working together with the *orbicularis oculi* – a muscle remote from cognition but intimately connected to – and illustrative of – 'the sweet emotions of the soul' (Ekman, 2004: 276). Another affective push begins to form: in perceiving these muscles, he enters an affective exchange – he can't not, he's pre-wired – and 500 microseconds later his own *zygomatic major* contracts. He doesn't think about it, it just happens 'without attention or conscious awareness' (Dimberg et al., 2000: 2), and as more microseconds pass, this almost imperceptible facial movement prompts neurochemicals associated with joy to flood his anterior cortical zones (Davidson et al., 1990). He reaches a 'resonant affective state'; their smiles have elicited his (Bänninger-Huber, 1992) and the prospect of volunteering begins to be something outside linear thought, something visceral, something felt.

The pushes are beginning to accumulate and he's moved to wonder at the practicalities of participating on the Programme: 'where? what? how?' To answer his questions, numerous vetted and authorised blogs allow volunteers to 'share their personal experiences of helping to eliminate extreme poverty across the developing world' (dfid.gov.uk). 'Eliminate', it's a grand promise that would ordinarily draw his scepticism but he's ready – *eager* – to be carried away. The first blog he comes to, 'Connecting Girls, Inspiring Futures', transports him to volunteer action in Afljalpur, Bangladesh: 'under the relentless sun . . . we begin our rally . . . shouting messages and carrying signs . . . we collected an even larger crowd as we went'. The blog leaps off the screen: 'everyone was in very high spirits by the time the rally finished . . . our final act was a dance by a 16-year-old girl who had been forced into marriage six days before' (dfid.gov.uk). The blog livelily promises him a collective and sensuous experience of rallied (and hot) bodies, passions, fears and hopes. As he reads on, these energies emerge and the crowd ascends to 'very high spirits' that succeed in 'connecting girls [and] inspiring futures'. His body is simultaneously charged with anticipation (of change) and hope (of a better future), and tempered by anxieties (towards forced marriage). Corporeally immersed in this world, he quickly loads the next linked-to blog that describes a commemoration for a Peruvian peace campaigner, Maria Elena Moyano. Again, the narrative enters him: Maria Elena's 'amazing work was sadly and brutally cut short . . . she was assassinated . . . for her determination to continue working for the community and her refusal to succumb to fear'. The commemoration is accordingly 'poignant' as a chant begins to build, 'Peruvians want peace, no more terrorism'; 'these words', says the blog, 'were particularly heartfelt'. The account closes: 'as I stood there, part of the crowd and part of something much greater than just myself, the words which really struck me were these: "You are Villa El Salvador. We are all Villa El Salvador. And Villa El Salvador will keep on progressing".' He reads through his body, placing himself in this richly emotional (sadness, poignancy, pride, hope, fear) scene and populated by a collective of (chanting) bodies whose intensities emerge in the volunteer's affected body: it is a 'poignant', 'heartfelt' and 'striking' ('struck') moment. Via these connections he follows the volunteer into a fleeting sense of transcendence: '[I'm] part of the crowd and part of something much greater than just myself.'[3]

This is a critical moment in his affective journey. The blogs enter him into a world replete with the 'buzz' of shared activity where moments are 'felt' through 'an energetic intensity of connection' (Conradson, 2003: 1987). And just as the volunteer writers feel a 'sense of belonging . . . more than [they] were a moment before', so too does he: 'this is the moment where world and individual, folded together, call each other into existence' (Dewsbury, 2003: 1910). The blogs therefore – in their vivid evocations of 'life' – take on an embodied presence and make him feel an affinity to folk who began as words but who emerge as affects. He wants to be there among them, experiencing all of the sensualities the rich descriptions implicitly promise. What began as another mundane day is now alive with the virtualities of a 'not-yet' that push on the actual of his present life choices (see Anderson, 2006). He dares to dream of something temporally and spatially excessive of his present. And it's a dream that's becoming more lucid; his passive wonder is slowly hardening his practical inquiry: 'what to gain? where to go?' The answer to these questions forms the final and decisive push.

He didn't know it but he has much to gain; he's part of a 'disillusioned youth' who lack ambition and aspiration. Though he hadn't thought of himself in these terms, they form a consistent theme whenever authority addresses him. The

Volunteering Programme exists because, the Prime Minister points out, his generation 'appear[s] lost. Their lives lack shape or any sense of direction. So they take out their frustrations and boredom on the world around them. They get involved with gangs. They smash up the neighbourhood. They turn to drink and drugs' (Cameron, 2010). It's a reality far away from his own (and that of many of his peers) but these words bring it closer, alarm him to a not-too-distant danger, demand evasive movement. The Programme offers a way out and thus becomes a move to stave off the stirring of anxiety in him: becoming a volunteer will 'kick start your career' and 'look great on your CV'. This is a familiar circulation of affects towards the (re)shaping of youth: aspiration rests on the body's disposition for transformation, sets off 'precognitive embodied impulses and feelings' that emerge 'as an affective orientation to the future' (Brown, 2011: 9). A 'spark of hope' flashes in him, pulses between synapses and a future possibility – a virtuality – begins to play out in the actuality of the present, steering his movements in the here and now (Anderson, 2006).

He's on the verge and clicks 'where can I go'; the website's response is to provide the final push. The webmaster has rendered the planet an irresistible sensorium. He's invited to 'choose a project' from '24 of the world's poorest countries' and presented with a range of maps and scenery-rich images (volunteerics.org). He's casually urged to 'have a browse through each of these country options' and is taken on a tour through an 'amazing rural community' in Bolivia, 'an idyllic setting of green bush lands' in Malawi and a version of South Africa 'bustling with hope and optimism for the future' (volunteerics.org). The maps put huge swathes of the Earth and an imaginary of seemingly infinite opportunities at his disposition. They conjure the globe as a wondrous 'dream space' (Tsing, 2005), carrying with it an 'imperative to imagine' new relations with new places and peoples (Jameson, 2005). There are 21 more of these evocative country profiles, each reproducing vibrant communities and sensuous landscapes. It so markedly contrasts with his study as to lure him: he's away, called by the affective 'hum of the world' (Bingham and Thrift, 2000: 281), pulled by an irresistible 'will to connect' (Ramsay, 2009) and the 'somatic internalisation' of distant places and peoples that brings with it the final and decisive affective imperative to 'become global' (Conradson and Latham, 2007: 233).

He clicks, checks the boxes, crosses his fingers. It's almost an autonomic reflex, his fingers obeying the will of his affected self and, by extension, the will of the champions of a Big Society populated by active citizens.

The rub

But here's the rub. The corporeal appeal of International Citizen Service draws him into a subject position idealised by modern (neo)liberal models of governance. Whatever his contribution to the 'fight against global poverty' in the Poor South, it shall last no more than ten weeks. After that time he's required to complete post-placement 'UK Action' in his home community: 'this final part of the programme demonstrates how the ICS volunteer experience can start a life long journey of positive social action at home' (volunteerics.org). His efforts to 'make a difference' and 'fight poverty' are now put into a wider context: 'your ICS journey doesn't suddenly stop when you complete your placement overseas', this is 'a new beginning' of a 'life of active citizenship'. The contribution to fighting global poverty, so important before, is now de-emphasised and he's refigured as a more 'productive' active citizen, one of those 'who contribute to the success of the free market economy by picking up the pieces that the free market drops: helping charities, clearing litter, helping inner city reconstruction' (Kellner, cited in Kinderman, 2011: 23). From this perspective the affective lure of ICS seems an elaborate ruse, more interested in fomenting domestic voluntarism than fighting the injustices of global poverty.

There is good reason to explore this journey with a readiness to believe in an affective push. It is a fact that our imagined subject's government trains a cohort of special policy makers to envisage a 'post-rational human subject' at whose 'mindspace' techniques of influence are to be aimed (Jones et al., 2011). He lives in an era of 'new brain-world' policy making in which the UK government (predominantly through the Behavioural Insights Team, or 'Nudge Unit') leads the way in research into 'emotional responses to words, images and events' that can be used to effect 'behavioural reactions' and 'changing minds' (Dolan et al., 2012). The Cabinet Office styles this as MINDSPACE policy making that 'use[s] the power of affect or emotion to stimulate behaviour change ... [and] to encourage other pro-social behaviours such as blood donation and community volunteering' (Cabinet Office, 2010: 30). There we have it: 'community volunteering', among other desirable behaviours, is an expressed target of its social marketing interventions (see Pykett et al., 2014). We must therefore take seriously the possibility that participants on the ICS volunteering programme are recruited into a life of active citizenship through a strongly affective appeal made directly to the body.

Researching affective worlds: active citizenship, conceptualisations and ethics

The imagining of pre-cognisant processes gives rise to significant ethical, conceptual and methodological challenges. For instance, in the broad sweep of tracking a passage of affect from policy to the body there is the implicit assumption that we are all, to borrow from Stuart Hall (1981), 'policy dopes' and this reduces social life to little more than a 'residual effect' of modes of governance (Barnett, 2005: 7). This highlights a failure to recognise our imagined subject as either a reflexive or affect*ing* actor; he's only ever a passive subject *being* affected. Within this there is also a narrow conceptualisation of affect. Even at its most empirical (in the field of clinical psychology, for instance), affect is understood as emergent from a 'raw domain of primitive experiential richness' that forms through 'transpersonal or pre-personal intensities' (Massumi, 2002: 29). As such, affects are essentially 'autonomous' (Massumi, 1995) and therefore an 'unstable object of governance' (Anderson, 2007: 162). We cannot then draw straight lines between affective experience and affective expressions of governance. This provides an important corrective, but only one that qualifies rather than contradicts – the body's autonomy does not prevent interested parties 'engineering' or 'manipulating' affects (Thrift, 2004; McCormack, 2008). The qualification, therefore, is that this is an account of the affective promotion of active citizenship; how bodies affect and are affected in the ebb and flow of sensory life remains to be seen.

To explore this ultimately unknowable world of affect, the data used here is largely visual and textual (for an account using ethnographic data from ICS volunteers see Griffiths, 2014). On visual methods, I followed a post-structuralist approach focused on the subjective impact a photograph may have on its viewer. Roland Barthes famously articulated this approach as the distinction between *studium*, a culturally informed interpretation of signs, and *punctum*, that visual property that escapes semiotic analysis to produce a more visceral impact; as Barthes describes it: '[that] which pricks me (but also bruises me, is poignant to me)' (2000: 27; see also Rose, 2007: 88–91). The conceptual premise of punctum, therefore, lends itself well to considering the affect that can be carried by imagery. An epistemologically similar approach was taken to the textual material used to promote ICS, concentrating on language that 'calls on the sensuous, the figurative, and the expressive' (Pelias, 2007: 183). This language formed emic codes that I link to the existing work on affect that is cited throughout the chapter. On writing, the style I use here begins from the notion that embodied experience is 'not easily told, but far easier felt' (Bennett, 2000: 120), attempting a form of 'performative writing' that aims to evoke 'worlds of memory, pleasure, sensation, imagination, affect, and in-sight' (Pollock, 1998: 80).

The result is a creative account of a network of marketers, webmasters, copywriters, politicians and civil servants, each contributing incrementally to an embodied appeal to participate in ICS. The cumulative push in evidence strongly suggests that in this case the promotion of active citizenship rests quite significantly on a specifically affective mode of emotional governance; on affect–emotion relations see Anderson (2006: 736–7) and Pile (2010: 9–10).

An epilogue

Months later he finds his time divided between (like a great many recent graduates) job applications, underpaid bar work and a local community programme run by Help the Aged. He visits Margaret three times a week to help with shopping, bureaucracy and general housekeeping. He enjoys it and he's happy that he discovered a pleasure in volunteering through his time in Malawi. He learnt a lot and it helped him become more aware of the world, and more confident in his ability to engage with it. He's most definitely active, a model of third sector efficacy. One Sunday, as he sets up the pill organiser, Margaret speaks of her own mother's old age: 'the nurse used to do that for her, *and* they used to send someone round to cook when she needed'. He knows that he's becoming more and more important. More support worker than volunteer. Margaret often speaks about 'them' and it's always about what *they* used to do. The mobile library, the community warden, the local baths; they're all now spoken of as pasts, as nostalgic figures of a better time. 'Still, at least . . .', she breaks off and looks up at her young helper with warmth: 'I don't know what I'd do without you'.

It's been a convoluted journey from that first contact when poverty touched him so. Those flooring statistics on dying children, the instinctual appeal of those smiling faces and the measures of hope and fear that finally pushed him to apply for the volunteering programme. Perhaps not every push emanated directly from the government, and certainly none of them was decisive, but slowly slowly his body was drawn to a practice of active citizenship. And they were right: his volunteering didn't stop at the end of his placement abroad. Here with Margaret he's content to provide in the gaps where now only nostalgia remains.

Notes

1 Full radio script available at: https://communication.cabinetoffice.gov.uk/fillers/wp-content/uploads/2013/10/ICS-Filler_RACC-Approval.pdf (accessed 20 July 2014).
2 A high quality stream of the advert is available at: www.youtube.com/watch?v=jmJFJZdS-D0 (last accessed 18 May 2014).
3 The two blogs are available at: https://dfid.blog.gov.uk/2012/03/14/connecting-girls-inspiring-futures/ and https://dfid.blog.gov.uk/2012/03/05/part-of-something-greater-than-myself/ (accessed 8 June 2014).

References

Ahmed, S. (2004) *The cultural politics of emotions*. Edinburgh University Press, Edinburgh.
Anderson, B. (2006) 'Becoming and being hopeful: towards a theory of affect', *Environment and planning D*, 24(5): 733–52.
Anderson, B. (2007) 'Hope for nanotechnology: anticipatory knowledge and the governance of affect', *Area*, 39(2): 156–65.
Anderson, B. (2012) 'Affect and biopower: towards a politics of life', *Transactions of the Institute of British Geographers*, 37: 28–43.
Anderson, B. and Adey, P. (2011) 'Affect and security: exercising emergency in UK civil contingencies', *Environment and Planning D: Society and Space*, 29: 1092–109.
Anderson, K. and Smith, S. (2001) 'Emotional geographies', *Transactions of the Institute of British Geographers*, 26: 7–10.
Bänninger-Huber, E. (1992) 'Prototypical affective microsequences in psychotherapeutic interaction', *Psychotherapy Research*, 2(4): 291–306.
Barnett, C. (2005) 'The consolations of "neoliberalism"', *Geoforum*, 36: 7–12.
Barnett, N. (2002) 'Including ourselves: New Labour and engagement with public services', *Management Decision*, 40(4): 310–17.
Barthes, R. (2000) *Camera Lucida: reflections on photography*. Vintage, London.
Bennett, K. (2000) 'Inter/viewing and inter/subjectivities: powerful performances', in Seymour, S., Hughes, A. and Morris, C. (eds) *Ethnography and rural research*. Countryside and Community Press, Cheltenham, pp. 120–35.
Bingham, N. and Thrift, N. (2000) 'Some new instructions for travelers: the geography of Bruno Latour and Michel Serres', in Crang, P. and Thrift, N. (eds) *Thinking space*. Routledge, London, pp. 281–301.

Brown, G. (2011) 'Emotional geographies of young people's aspirations for adult life', *Children's Geographies*, 9(1): 7–22.

Cabinet Office (2010) 'MINDSPACE: Influencing behaviour through public policy', at: www.instituteforgovernment.org.uk/our-work/better-policy-making/mindspace-behavioural-economics (accessed 8 June 2014).

Cameron, D. (2010) 'Foreword', *National Citizen Service Policy Paper*. Conservative Party, London.

Cameron, D. (2011) 'Why we're right to ringfence the aid budget', *The Guardian*, 11 June 2011, at:www.guardian.co.uk/global-development/2011/jun/11/david-cameron-defends-aid-funding (accessed 28 June 2014).

Clarke, J., Newman, J., Smith, N., Vidler, E. and Westmarland, L. (2007) *Creating citizen-consumers: changing publics and changing public services*. Sage, London.

Connolly, W. (2002) *Neuropolitics: thinking, culture, speed* (Vol. 23). University of Minnesota Press, Minneapolis.

Conradson, D. (2003) 'Doing organisational space: practices of voluntary welfare in the city', *Environment and Planning A*, 35(11): 1975–92.

Conradson, D. and Latham, A. (2007) 'The affective possibilities of London: antipodean transnationals and the overseas experience', *Mobilities*, 2(2): 231–54.

Conservative Party (2009a) National Citizen Service, at: www.conservatives.com/News/News_stories/2010/04/Conservatives_launch_plans_for_a_National_Citizen_Service.aspx (accessed 8 June 2014).

Conservative Party (2009b) One World Conservatism: a Conservative agenda for international development. *Policy Green Paper* no. 11, at: www.conservatives.com/~/media/Files/.../OneWorldConservatism (accessed 8 May 2014).

Davidson, R.J., Ekman, P., Saron, C.D., Senulis, J.A. and Friesen, W. (1990) 'Approach-withdrawal and cerebral asymmetry: emotional expression and brain physiology: I', *Journal of Personality and Social Psychology*, 58(2): 330–41.

Dewsbury, J.D. (2003) 'Witnessing space: "knowing without contemplation"', *Environment and Planning A*, 35: 1907–32.

Dewsbury, J.D. (2010) 'Performative, non-representational, and affect-based research: seven injunctions', in DeLyser, D., Herbert, S., Aitken, S., Crang, M. and McDowell, L. (eds) *The Sage handbook of qualitative geography*. Sage, London, pp. 321–34.

Dewsbury, J.D., Harrison, P., Rose, M. and Wylie, J. (2002) 'Enacting geographies', *Geoforum*, 33(4): 437–40.

Dimberg, U., Thunberg, M. and Elmehed, K. (2000) 'Unconscious facial reactions to emotional facial expressions', *Psychological Science*, 11(1): 86–9.

Dolan, P., Hallsworth, M., Halpern, D., King, D., Metcalfe, R. and Vlaev, I. (2012) 'Influencing behaviour: the mindspace way', *Journal of Economic Psychology*, 33(1): 264–77.

Ekman, P. (2004) *Emotions revealed: understanding faces and feelings*. Orion, London.

Fyfe, N. (1995) 'Law and order policy and the spaces of citizenship in contemporary Britain', *Political Geography*, 14(2): 177–89.

Griffiths, M. (2014) 'Transcending neoliberalism in international volunteering', in Dashper, K. (ed.) *Rural tourism: an international perspective*. Cambridge Scholars, Newcastle, pp. 115–33.

Guardian (2013) 'The demise of Choice FM', 14 November 2013, at: www.theguardian.com/commentisfree/2013/nov/14/demise-choice-fm-pirate-radio-black-britons-capital-xtra (accessed 15 July 2014).

Hall, S. (1981) 'Notes on deconstructing "the Popular"', in Szeman, I. and Kaposy, T. (eds) (2010) *Cultural theory: an anthology*. Wiley, London, pp. 72–80.

Haste, H. (2004) 'Constructing the citizen', *Political Psychology*, 25(3): 413–39.

Hook, D. (2007) *Foucault, psychology, and the analytics of power*. Macmillan, London.

Isin, E. (2004) 'The neurotic citizen', *Citizenship Studies*, 8(3): 217–35.

Jameson, F. (2005) *Archaeologies of the future: the desire called utopia and other science fictions*. Verso, London.

Joffe, H. (2008) 'The power of visual material: persuasion, emotion and identification', *Diogenes*, 55(1): 84–93.

Jones, R., Pykett, J. and Whitehead, M. (2011) 'The geographies of soft paternalism in the UK: the rise of the avuncular state and changing behaviour after neoliberalism', *Geography Compass*, 5(1): 50–62.

Katz, J. (1999) *How emotions work*. University of Chicago Press, Chicago, IL.

Kearns, J. (1992) 'Active citizenship and urban governance', *Transactions of the Institute of British Geographers*, 17(1): 20–34.

Kinderman, D. (2011) '"Free us up so we can be responsible!" The co-evolution of corporate social responsibility and neo-liberalism in the UK, 1977–2010', *Socio-Economic Review*, 10(1): 1–23.

Lousley, C. (2014) '"With Love from Band Aid": sentimental exchange, affective economies, and popular globalism', *Emotion, Space and Society*, 10: 7–17.

McCormack, D. (2008) 'Engineering affective atmospheres on the moving geographies of the 1897 Andrée expedition', *Cultural Geographies*, 15(4): 413–30.

Massumi, B. (1995) 'The autonomy of affect', *Cultural Critique*, 31: 83–109.

Massumi, B. (2002) 'Navigating movements – with Brian Massumi', in Zournazi, M. (ed.) *Hope: new philosophies for change*. Pluto Press, Sydney, pp. 210–44.

Massumi, B. (2005) 'Fear (the spectrum said)', *Positions: East Asia Cultures Critique*, 13(1): 31–48.

Pelias, R. (2007) 'Performative writing: the ethics of representation in form and body', in Denzin, N. and Giardina, M. (eds) *Ethical futures in qualitative research: decolonizing the politics of knowledge*. Left Coast Press, Walnut Creek, CA, pp. 181–96.

Pile, S. (2010) 'Emotions and affect in recent human geography', *Transactions of the Institute of British Geographers*, 35(1): 5–20.

Pollock, D. (1998) 'Performing writing', in Phelan, P. and Lane, J. (eds) *The ends of performance*. New York University Press, New York, pp. 73–103.

Poole, L. (2000) 'Health care: New Labour's NHS', in Clarke, J., Gewirtz, S. and McLaughlin, E. (eds) *New managerialism, new welfare?* Sage, London, pp. 102–21.

Pykett, J., Jones, R., Welsh, M. and Whitehead, M. (2014) 'The art of choosing and the politics of social marketing', *Policy Studies*, 35(2): 97–114.

RACC (Radio Advertising Clearance Centre) (2013) 'Script clearance form for ICS radio advertisement', at: https://communication.cabinetoffice.gov.uk/fillers/wp-content/uploads/2013/10/ICS-Filler_RACC-Approval.pdf (accessed 20 July 2014).

Ramsay, N. (2009) 'Taking-place: refracted enchantment and the habitual spaces of the tourist souvenir', *Social and Cultural Geography*, 10(2): 197–217.

Rose, G. (2007) *Visual methodologies: an introduction to researching with visual materials*. Sage, London.

Rose, N. (1996) 'Governing "advanced liberal democracies"', in Barry, A., Osbourne, T. and Rose, N. (eds) *Foucault and political reason*. UCL Press, London, pp. 37–64.

Ross, A. (2007) 'Multiple identities and education for active citizenship', *British Journal of Educational Studies*, 55(3): 286–303.

Sharpe, J., Campbell, P. and Laurie, E. (2010) 'The violence of aid? Giving, power and active subjects in One World Conservatism', *Third World Quarterly*, 31(7): 1125–43.

Thrift, N. (2004) 'Intensities of feeling: towards a spatial politics of affect', *Geografiska Annaler*, 86B: 57–78.

Thrift, N. (2007) *Non-representational theory: space, politics, affect*. Routledge, London.

Tsing, A. (2005) *Friction: an ethnography of global connection*. Princeton University Press, Princeton, NJ.

15 The relational spaces of mentoring with young people 'at risk'

Fiona M. Smith, Matej Blazek,
Donna Marie Brown and
Lorraine van Blerk

Introduction

In this chapter we examine the relational spaces of emotional work in a mentoring project which deployed volunteer mentors ('active citizens') to complement formal structures of state engagement with vulnerable young people deemed 'at risk' of anti-social and criminal behaviour.[1] In so doing, we explore the complexities of the kind of emotional work involved in policy-in-practice, particularly in policy interventions which might at a general level be critiqued as representing individualising neoliberal modes of governance which 'responsibilise', or even stigmatise, individuals (Bowlby et al., 2014; Pykett, 2014). Critics argue that such policies target attention on the need to discipline what are viewed as problematic emotions and related behaviours (a particular characteristic of many policy interventions with young people, Kraftl and Blazek, 2015), while failing to address wider structural inequalities. However, by looking more closely at how emotions are embedded in wider relational practices of care, we examine how those who participated both valued the emotional labour involved and insisted on the need to address some of the limitations of such models of practice. This in turn engages with wider discussions (Newman, Chapter 2, this volume; Laurie and Bondi, 2005) on exploring the risks and opportunities of the apparent co-option of emotional work into the emotional, neoliberal state by refusing any simple application of the somewhat totalising logics of neoliberalism. It instead demonstrates how other rationales and modes of practice may insist on the potential for other forms of emotional practice to emerge. This includes the recognition both of young people's own embodied emotional agencies and of the need for supportive structures and relations of care alongside approaches which insist on the need to address wider aspects of inequality and exclusion.

The chapter proceeds by outlining the nature of the project and situating it in relation to wider policies of 'early intervention' with young people. We develop understandings of the relational spaces of care, drawing on Bondi's (2008) discussion of the relational theory of practice which emphasises interpersonal relationships and dynamics between service providers and their clients as not just contingencies but as the core mediums of policy delivery (Hunter, 2012). Following debates about young people's agency within and outside the neoliberal mainstreams of both the Global North and South (Evans, 2012; Punch and Sugden, 2013; Blazek et al., 2015), we focus on viewing young people not only as 'recipients' of care (Wiles, 2011) but also as active participants in the relational practices of policy delivery (Dickens and Lonie, 2013). Thus not only are the mentors and project managers participants in the emotional work of mentoring but the young people are too, alongside others in the wider networks of practice. After briefly describing the methodologies deployed in the evaluation, we examine the relational practices of the mentoring project, which involved a complex mix of more obviously 'emotional' aspects intertwined with more explicitly practical, situated activities and engagements which we argue together constitute the emotional work of mentoring. We then outline the embeddedness of these practices in wider relational networks of care and identify how the emotional work of those involved simultaneously valued the work of the project and engaged with its limitations. We therefore seek to make space, on the one hand, to acknowledge the complex impacts of neoliberalisation (and more recently austerity) in the ways that social policy practice is framed and reframed in diverse forms of practice, while simultaneously arguing that a focus on the relational nature of emotional work might leave space to understand how practices of social justice (Griffiths, 2013), other models of practice, both professional and lay, as well as the agency of the young people themselves may exceed the narrow bounds of neoliberal norms and emerge within the complex spaces of care in work (both professional and voluntary) with young people (Blazek and Windram-Geddes, 2013; Blazek and Kraftl, 2015).

Relational practices of social policy in youth mentoring

Young people's lives have increasingly become the focus of policy interventions. Within the repertoire available, mentoring[2] is a widespread model of practice (Rhodes and Lowe, 2008; Du Bois et al., 2002). It can be deployed on issues ranging from informal learning to schooling (Pryce, 2012; Sandford et al., 2010) and is often targeted at young people experiencing social disadvantage or deemed 'at risk' (Moodie and Fisher, 2009). While many such programmes develop within the third sector, the project we examined, while having a precursor in the voluntary sector partner organisation's own work, is an example of the multi-agency programmes working across state, third and (sometimes, though not in this case) private sectors which have emerged as part of what Jupp (2013) has called the 'thickening' of social policy interventions targeting particular population groups. As these multi-agency strategies are mobilised to develop forms of governance around the behaviours of young people, practices of mentoring are situated within wider circuits of social policy.

The specific project discussed in this chapter was a youth mentoring project, *plusone mentoring* in Scotland (UK), which used volunteer adult mentors trained and managed through an established youth work voluntary organisation (YMCA) to work within a multi-agency partnership model with young people deemed to be at risk of offending or anti-social behaviour (Blazek et al., 2011). It was focused primarily on aspects of youth justice policy and practice and reflected evidence from policy reviews which emphasise the value of targeted preventative and 'pro-social' programmes in decreasing such risk, particularly in the early teenage years (Greenwood, 2008; McAra and McVie, 2010). It fits directly with the centrality of a focus on 'early intervention' in the devolved administration of Scotland in the fields of education, social work and criminal justice under the policy known as 'GIRFEC' – 'Getting it right for every child' (Scottish Government, 2008a, 2008b). The project discussed here sits within stated national priorities across diverse aspects of policy related to young people (Croall, 2006; Scottish Government, 2009, 2013; Scottish Parliament, 2014: Education Scotland, 2014: Sercombe, 2009; Scotland's Commissioner for Children and Young People, 2014). In youth justice, mentoring and 'diversion' for specific young people sit alongside broader preventative measures, including the Children's Hearings system, efforts directed at building community capacities and utilising community wardens (Allen and Stern, 2007; Brown, 2013; Children's Hearings Scotland, 2015), as well as more 'carceral' approaches (Schliehe, 2014) such as secure care units or prison (for those over 16), within a complex arrangement of diverse criminal justice bodies and partnerships (Audit Scotland, 2011). Early intervention and prevention approaches tend to emphasise working across agencies and different sectors, with models of practice from the voluntary sector also being examined as having potential, for example, to access otherwise difficult-to-reach groups, who may be much less likely to engage with statutory agencies such as the police or social work services.

Such 'new models' of working are also part of an agenda of public service reform which sits within a wider framework of decreasing funding for public services and which has included such diverse moves as increasing centralisation (in the case of policing and fire and rescue services) and increasing tendencies to 'off-load' what are seen as 'soft services' (with the implied devaluing of more emotional and caring roles) to lower cost, non-state provision, such as third sector providers (Bunt and Harris, 2010; Hamnett, 2014), although one might also point to some aspects of different political discourse in the devolved administration in Scotland (Law and Mooney, 2012). Thus it might be tempting to cite projects such as the one in this chapter as an example of increasingly neoliberal state structures in Scotland, the wider UK and elsewhere which are deploying non-state agencies, including the voluntary sector and voluntary labour, to enact practices of 'responsibilising' emotional governance with those whose behaviour is deemed problematic (Bowlby et al., 2014) in the context of a diminution of the state in times of austerity (Clayton et al., 2015). At the same time, the targeting of measures might stigmatise some young people and overlook others (and their needs) by stepping away from 'universalism' towards targeted provision, as youth work and youth justice fuse (Williamson, 2009) and austerity measures contest (or even undermine) the diversity of professional youth work provision (Bradford and Cullen, 2014). In response, Tiffany (2012) calls for '"targeting through universalism" – making youth work and support available to all but having an eye for those who need it most'.

In order not to read off the meaning and experience of the mentoring process from such structural features and to attend to the relationality of the emotional work of mentoring in practice, however, we emphasise here the lived experiences of the doing of social policy (Jupp, 2008). Following Smith et al. (2010: 270), we argue for the importance of attending to the 'situated, emotional and embodied' nature of social policy as well as the importance of the 'more-than-social' in exploring 'how the spaces of [social practice] function in and through myriad prosaic, complex, tangible and intangible practices, feelings and encounters'. We argue this helps us to engage critically with making the practices of policy in action 'more real and more credible as objects of policy and activism' (Gibson-Graham, 2008: 613; see also Askins, 2015; Conradson, 2003; Jupp, 2013) and to examine how they are central to the relational practices of care.

An explicitly 'relational' approach to analysis of social policy and practice (Hunter, 2012) attends to more unexpected, more contingent notions of the effects of policy, reflecting Horton and Kraftl's (2005) argument that usefulness arises in practice. It emphasises the views and involvement of 'recipients' (as well as providers) of care and focuses on the emotional (that is, intersubjective) and relationship work, 'however ordinary', through which care is enacted (Bondi, 2008: 262). However, we also embed these immediate relational practices in wider relational networks, considering how the socio-material geographies of young people's lives and their (not always unproblematic) relationships to their families, 'communities', institutional support and links from mentoring to broader multi-agency interventions all potentially impact on their experience of the mentoring project, arguing that these emphatically social networks are also 'agents of care' (Gibson-Graham, 2008).

Methodologies in the research with *plusone mentoring*

plusone mentoring, which was launched in September 2009 in three pilot areas in Scotland, aimed to develop an early intervention approach by offering mentoring to young people (mainly aged 8–14) identified as being at high risk of offending or anti-social behaviour, using volunteer adult mentors trained and managed through local YMCA centres and working in a multi-agency partnership. The Violence Reduction Unit (VRU) of the Scottish Police and the three local authorities in which the project was introduced provided two years of funding from September 2009. The project's Oversight Group consisted of representatives of the VRU, the Association of Directors of Social Work in Scotland, and YMCA Scotland. The project has subsequently developed and expanded to cover, by 2015, ten local areas in Scotland, but we focus here only on evidence from the Scottish Government-funded evaluation of the first phase of the project, which we conducted from January to May 2011 (Blazek et al., 2011).

Once assessed as 'at risk' in relation to a number of factors (such as parenting difficulties, existing hostile or violent behaviour, criminal or anti-social behaviour, and substance misuse), young people were referred to the project by local multi-agency panels, consisting of representatives of the police, social work, education, community mental health teams and others. They were then offered one-on-one mentoring by volunteers who were recruited, trained, supervised and supported by local YMCA-based project managers. Participation by young people and their families was entirely voluntary. If they agreed, the young person was matched with an adult volunteer mentor who then met them for one session per week.[3] Around eighty mentors were trained in this first phase by the YMCA. Training emphasised the youth work ethos of the project, which utilised a young person-centred youth work approach drawing on common principles such as 'young people choose to participate; the work builds from where young people are and the young person and youth worker are partners in the learning process' (Education Scotland, 2014: 4). There was stress on the need to offer long-term commitment to the young person over a timescale of a year or more (addressing critiques about the short-term nature of many interventions – Grossman and Rhodes, 2002; Judge, 2015), and the aim was ultimately to enable the young person to become independent of the mentoring process.

Evaluation of the project utilised a mixed methodology that sought to put the young people's perspectives at the centre of the research (Barker, 2008; van Blerk and Kesby, 2009). It involved semi-structured interviews with young people, largely in informal settings where the young person would normally have been mentored or in the YMCA centres, and interviews with adult volunteer mentors, the three programme managers and with those on the Oversight Group and local Referral Groups. Relevant documentation was also reviewed, including referral forms for the young people, reports of mentoring sessions, training materials for mentors and the policy materials from local Referral Groups. The research received ethical clearance from the University of Dundee Research Ethics Committee and was developed in accordance with ethical guidelines for working with young people (Alderson and Morrow, 2011), using informed consent and age-appropriate information for the participants. We do not name here any of the pilot areas, but they were publicised by the project in their own materials. As a result we omit any details that would allow individual young people or their mentors to be identified, which means that some potentially useful contextual material (for example the age or gender of the young person) is absent. The tight timetable of the evaluation and the need to avoid disruption to the often fragile process of developing the mentoring relationship meant techniques such as ethnography or more participatory methods were not adopted.

Participating in the emotional work of the mentoring relationship

From the outset volunteer mentors were trained to develop a relationship with the young person they mentored based on key youth-work related principles (partnership with the young person, the young person's voluntary participation, progressive empowerment, an informal and friendly atmosphere). This relational work was seen as connecting to three phases of mentoring – 'The beginning: developing rapport and building trust'; 'Developing the relationship: working together to reach goals'; 'Ending, re-defining and evaluating' (Mentor training pack). Thus from the outset we see a complex mix of emphasis on more explicitly 'emotional' aspects ('rapport', 'trust'), the work of developing a relationship ('working together') and a more strategic sense of 'goals' (though they are ones to be developed in partnership with the young person). Programme managers talked about the importance of getting the 'match' between mentor and mentee right from the start. This might include aspects of similarity in a demographic sense, but more typically whether they would 'get on' or had shared interests.

Focusing on the importance of the work in the first few weeks, young people and mentors talked about an initial 'breakthrough' as the relationship began to evolve. There were discussions between both on how the relationship might change, often involving the young person being more willing to engage with the mentor, even if, in the context of the complex set of issues some of the young people faced, these changes might be relatively small-scale (a young person being willing to talk a bit more, to look the mentor in the eye as they talked, or showing signs of improved personal care, such as having brushed their teeth). Thus any 'breakthrough' moment could be, as mentors and project managers stressed, something very undramatic, although some young people also mentioned that the changes were apparent to themselves or to others such as their teachers or families. Mentors and project managers emphasised the need for sustained engagement and 'patience':

> You need to be patient. You need to be committed. You cannot judge but must try to understand instead . . . That's how you can make a difference with the young person.
>
> (Mentor interview)

The process was designed to develop this patient engagement and the ongoing participation of the young person in the second stage of mentoring which, using an informal and friendly atmosphere, aimed to help the young person be able to identify goals for themselves and to address some often quite practical ways they and the mentor might 'work together' to achieve them. Despite formal definitions of mentoring as a 'systemic intervention' (Keller, 2005) and the multi-agency structures through which young people were referred to the project, few talked in any explicit way about the programme as an 'intervention' scheme. Although some family members mentioned this, the young people instead talked about their mentors 'helping them', mentioning aspects such as emotional problems, social relationships or educational issues and stressing the importance of having an adult who would 'listen', would be 'here for me' and would support them in a non-judgemental way.

Mentors were regarded as different from parents or other adult family members (though some saw similarities with relationships to people such as their social workers) and the young people sometimes referred to them as 'friends'. Other studies confirm that 'friendship' is a key way young people make sense of the mentoring relationship (Philip et al., 2004), while Askins (2015) reports how adults engaged in a 'befriending' scheme very quickly moved towards using the term of 'friends' rather than 'befrienders/befriendees' to describe their relationship. However, the young people were also aware that this was not the same as other friendships – sometimes being very positive that it was 'different' from their friendships with their peer groups, mentioning the value of 'mature' intergenerational support. Added to this was the importance of the mentor being reliable and trustworthy, particularly since a number of the young people stated that they had few friends and/or felt social or emotional support was missing in their home life, and given the wider vulnerabilities affecting the young people on the programme and which were often key factors in their referral to it. At the same time, the mentoring project had clear points of difference from non-mentoring based friendships and although this was a 'voluntary relationship' which the young people were free to end at any time (Bowlby, 2011: 607), which did sometimes happen, it was mediated by a prior commitment by the volunteer mentor to be willing to undertake mentoring for normally at least twelve months, something that was explained to the young people and their families at the outset. Thus the sense of trying to find a suitable way to describe the relationship with the mentor – like a social worker, like a friend, but an adult and not like their peers – is perhaps an accurate description of the distinctive and negotiated nature of the relationship offered by mentoring for the young people. To support this, and in recognition of the particular nature of that relational encounter, there was ongoing close support and supervision by local programme managers, and emphasis on the need for adherence to child protection practices and to appropriate forms of conduct with the young people.

It is clear, therefore, that the 'remarkable and the unremarkable' (Meth, 2008: 41) were present in the emotional work of the mentoring relationship which formed a core element in how the social policy 'intervention' might be delivered. However, this relational work was also fundamentally practical and active, engaging young people in what youth justice approaches term 'diversionary activities' (McAra and McVie, 2010) and what others might describe as a type of informal education (Mills and Kraftl, 2014). Mentors were given clear guidance that they were not qualified to offer behavioural counselling or other therapeutic roles. Rather, changes in emotional, behavioural and social skills were to be developed through practical approaches and embodied experience, where the mentoring relationship was at the centre of a number of spaces and networks of support for the young people. Mentoring explicitly happened outside the young person's home space, in sites which they often already knew (football pitches, cafes, parks, leisure spaces and, rarely, the YMCA centres) and were usually suggested by the young person. The relational process of mentoring drew such ordinary sites and spaces into the emotional work of the project as informal 'transitory spaces of care' (Johnsen et al., 2005: 323), using them as sites of 'co-presence' in which different practical activities could take place while the face-to-face, embodied meetings facilitated the development of the mentoring relationship. These places together with 'seemingly mundane acts' (Staeheli et al., 2012: 630) – having a chat, playing football, going for a walk – were central to the practical interventions of the project.

The high incidence of chaotic or problematic family circumstances among the assessed risk factors which initially led to the young people being referred to the project meant being out of their everyday spaces was viewed positively by some of the young people. On a practical level, being involved with their mentor was also a key means by which the young people might access activities and spaces that they might otherwise find difficult to access due to an absence of opportunities for leisure-time activities in their home area, meaning the activities with their mentor were their only options. Alternatively, young people might have been unable to access activities due to lack of confidence, lack of friends to go with them, tight family finances or limited family support. Those for whom this was particularly the case were more likely to make suggestions that mentoring should happen more frequently than once a week. In contrast, those who valued more the emotional-relational aspect of the mentoring, or who placed more emphasis on using the mentoring process to find ways of addressing particular problems, were less likely to ask for additional sessions, although the distinction was not a clear-cut one.

All of the young people reported some value in the activities they undertook, but there was a sense from some that, while they saw benefits, they also recognised the limitations of the scheme, particularly because, despite perhaps increasing their own personal resilience, the mentoring scheme itself could not address wider structural problems: 'in our neighbourhood . . . there is nothing else you can do there, nowhere to go' (interview with young person). Clearly the young person articulates here what is a more general critique of mentoring and other interventions focused primarily on the individual, namely that while the young people may be removed temporally from their everyday environment during the mentoring process (and may value this), wider structural problems such as poor public service provision and the conditions which lead to anti-social behaviour are outside the scope of the intervention (Tiffany, 2012). But it is perhaps too tempting, as critical social science researchers, to dismiss the potential value of such work and to miss careful attention to the relational emotional work developed in this kind of practice which many of the young people indicated they felt to have been helpful in at least some ways, and it is important not to dismiss the potential value of more individually focused forms of social policy, as expressed by the young people, in the context of wider critiques. Furthermore, when analysing how the relational work of mentoring progressed in the project, we would argue that a narrow focus on the individual one-on-one encounters also misses the significance of wider networks of care which are important in the overall emotional work of the project.

Embedding mentoring in wider relational networks of care

Recognising wider networks of care is important for understanding the situatedness of practices of care, and of mentoring in particular, and moving away from a close focus on the direct mentor–mentee relationship, significant though this may be (Keller, 2005; Bowlby, 2011). Asked to outline what they expected when they signed up for mentoring, one young person referred to the fact that it was mainly their mum who thought about it and that the role of the programme manager from the local centre coming to their house to explain was important. That everyday management of the project was based in the local YMCA centres made some difference to whether the families of the young people referred to the project were open to them taking part as it was perceived to have a more positive image among some, particularly compared to what were reported as negative experiences of some statutory agencies (police, schools or social work, for example, though by no means uniformly). Although opinions varied considerably, the YMCA was seen by some families and young people as providing opportunities rather than being an agency which could apply sanctions to the young people and their families. At the same time, although this points to the positive contribution of non-statutory/voluntary section organisations in providing this kind of care, rather than foregrounding any voluntary/paid work binary it is worth noting the key role of local (paid) project managers (based in the YMCA) in facilitating and maintaining the mentoring process through developing other relationships. In particular, relationships with the young people's families were very important, especially since 'parenting difficulties' were themselves the factor with the highest average 'risk' in the assessments used for referral to the project (Blazek et al., 2011: 20–1). Also, despite the mentoring taking place outside their homes, families and other aspects of the young people's everyday lives were a key factor in the emotional, relational work of mentoring. This might include seeing that mentoring might alter family relationships ('learning how to talk to my mum'), situations where other siblings asked if they could get mentoring too (even some-times threatening to misbehave if they did not) and evidence from the schools in one area which reported that the young people were not experiencing stigma for their participation but rather envy from their peers at getting to participate in the activities.

These wider connections were, however, not incidental to the process (although often the specific outcomes were not directly intended ones). Evidence from the referral process, mentors, programme managers and interviews with the young people themselves indicated that many of the young people who were referred struggled in everyday social relationships, including family, school and com-munity, as well as in relationships with other professionals and statutory agencies. Therefore, a central element of the mentoring relationship was to provide practical advice and help to mediate and facilitate engagement between the young person and other institutions or groups (school or college, sports clubs, and so on). This could involve efforts from the volunteers beyond the one-hour mentoring sessions, for example accompanying a young person on the way to college for the first few weeks of their studies. Mentors and programme managers also mediated and networked with other institutions, such as a young person's school after they had been expelled or had dropped out due to bullying, or helping a young person wrongly accused of anti-social behaviour in their community by their peers in order to develop a way to counteract the accusations. In this context, the role of the mentor (and programme manager) as people with more 'formal' positions in relation to the young person could be key, as could the reputation and position of the YMCA within the local multi-agency partnerships. Thus aspects of the mentoring programme explicitly recognised the need to work to facilitate connections and engagements between the young person and a range of spaces and activities beyond the more 'transitory' but recurring spaces of mentoring.

In terms of thinking about mentoring as a process in which emotional work develops in reciprocal forms, i.e. not only as characterised by 'asymmetric' relationships (Korf, 2007) in which the adult mentor provides care and the young person is the recipient – although the power relations of mentoring and other forms of care need to be kept clearly in mind (Bondi, 2008) – we can also consider how those such as the volunteer mentors might benefit from the mentoring process too. In that way recipients of care might not only be figured as 'vulnerable' in the sense of 'fragility or weakness' (Wiles, 2011: 573); rather they might be viewed as bringing their own contributions to the mentoring process too, emphasising ideas about the active participation and emotional work involved for all those in the mentoring process. Interestingly, the project explicitly set out to foreground that there would be benefits for volunteers that, in typical policy terms, might 'build capacity' in their communities (though this was perhaps rather loosely specified). Mentors were slightly more specific about what they felt they gained. For some it was experience to equip them for future career development in fields such as social work, education or community and youth work. Thus skills, experience and understanding developed within the relational practices of mentoring might inform future work (either paid or voluntary). Others emphasised that their motivations were driven more by altruism ('giving back') or commitments to social justice ('doing something for the young people').

Across all mentors, however, there was evidence not only that they entered the mentoring experience with particular motivations, in some cases relatively strategic ones, but also that engagement in the project had developed their understanding of and their capacity to work with young people at risk (Roberts and Devine, 2004), including in ways which would not otherwise have been accessible to them through training or their own life experiences: 'I began to see better what some young people experience and especially how incredibly difficult some of those things they encounter are. It's something I have not experienced myself, something I'm not sure how I would've responded to' (interview with mentor). Thus it becomes possible to consider how the experience may be one which is at least potentially transformative in some ways for the adult volunteers and not only for the young people. While any parallel with the reciprocal relations of friendship would be overdrawn, this does offer opportunities for a 'transformative politics of encounter' (Askins, 2015: 473) which exist (even) within practices which also sit within wider networks of the neoliberal state and which suggest possibilities for alternative and diverse modes of emotional work.

That said, it remains important to consider how the immediate and proximate emotional and relational work of mentoring remains situated within the broader aims of programme providers and statutory funders, and, ultimately, national-scale frameworks of youth justice and child protection. The extent to which young people were aware of this was not explored in the evaluation and would be an interesting question in future research. There also remains scope for future consideration of how young people's engagements in such projects might be 'up-scaled' to think about how schemes such as this 'implicate' young people within wider political structures and relations (Hopkins and Alexander, 2010). At a more local scale, there is some evidence of the possibilities of forming connections between changes at the local/interpersonal level and wider social processes. Brown's (2013) study of community wardens, for example, has examples of where attitudes towards young people had changed in their local communities. However, the scaling up between the individual and wider issues remains a key tension of such work.

There are also questions about how efforts to support young people and build capacity and skills in working with young people in a project such as this are sustained into longer-term change for the young people, the volunteers and in relation to wider policy goals. This is particularly the case in the context of sometimes enabling but often (increasingly) challenging institutional and financial conditions under which social policies and practices are being reshaped in conditions of neoliberalism and austerity (Clayton et al., 2015). We acknowledge the real limitations of such programmes in relation to wider structural inequalities in the availability of and access to facilities by young people, the ways ongoing (and often intensifying) impacts of deprivation and austerity affect this, and the constant need to examine the power relations which underpin and shape interventions in the lives of young people. Likewise, attending closely to young people's individual experiences still entails the risk of either decontextualising them or seeing them as 'the prompt for [only] a particular policy or practical intervention' (Kraftl and Blazek, 2015: 297). However, we have sought here to outline a wider relational approach to the emotional work of social policy-in-practice. In so doing, we concur with arguments that (re)connecting young people's lives with the realm of social policy requires a shift from considering the 'outcomes' of 'interventions' towards understanding young people as neither the 'recipients' of care nor 'problems' to be solved, but rather as active participants co-constructing the relational emotional work of policy-in-practice.

Notes

1 We would like to thank the local project managers for their help in undertaking this research, and all the young people and volunteer mentors who participated in the study. The Scottish Government funded the evaluation study on which the chapter is based.
2 We use here the term 'mentoring' as it is used by the project evaluated. While terms such as 'befriending' and 'buddying' are often deployed relatively interchangeably in practice, the specific nature of the relationships developed and the approaches to intervention/ social care vary considerably even where similar terms are used.
3 Some referrals were deemed inappropriate for *plusone mentoring* and were passed to other agencies and in thirteen of the ninety-six referrals in this phase the young people or their families declined the offer of mentoring.

References

Alderson, P. and Morrow, V. (2011) *The ethics of research with children and young people: a practical handbook.* Sage, London.

Allen, R. and Stern, V. (eds) (2007) *Justice reinvestment: a new approach to crime and justice.* International Centre for Prison Studies, King's College London, London.

Askins, K. (2015) 'Being together: everyday geographies and the quiet politics of belonging', *ACME: An International E-Journal for Critical Geographies*, 14(2): 470–78.

Audit Scotland (2011) *An overview of Scotland's criminal justice system.* Audit Scotland, Edinburgh.

Barker, J. (2008) 'Methodologies for change? A critique of applied research in children's geographies', *Children's Geographies*, 6: 183–94.

Blazek, M. and Kraftl, P. (eds) (2015) *Children's emotions in policy and practice: mapping and making spaces of childhood.* Palgrave Macmillan, London.

Blazek, M. and Windram-Geddes, M. (2013) 'Editorial: thinking and doing children's emotional geographies', *Emotion, Space and Society*, 9: 1–3.

Blazek, M., Brown, D.M., Smith, F.M. and van Blerk, L. (2011) *plusone mentoring evaluation.* University of Dundee/Scottish Institute of Policing Research, Dundee.

Blazek, M., Smith, F.M., Lemešová, M. and Hricová, P. (2015) 'Ethics of care across professional and everyday positionalities: the (un)expected impacts of participatory video with young female carers in Slovakia', *Geoforum*, 61: 45–55.

Bondi, L. (2008) 'On the relational dynamics of caring: a psychotherapeutic approach to emotional and power dimensions of women's care work', *Gender, Place and Culture*, 15(3): 249–65.

Bowlby, S. (2011) 'Friendship, co-presence and care: neglected spaces', *Social and Cultural Geography*, 12(6): 605–22.

Bowlby, S., Lea, J. and Holt, L. (2014) 'Learning how to behave in school: a study of the experiences of children and young people with socio-emotional differences', in Mills, S. and Kraftl, P. (eds) *Informal education, childhood and youth: geographies, histories, practices.* Palgrave, London, pp. 124–39.

Bradford, S. and Cullen, F. (2014) 'Positive for youth work? Contested terrains of professional youth work in austerity England', *International Journal of Adolescence and Youth*, 19: 93–106.

Brown, D.M. (2013) 'Young people, anti-social behaviour and public space: the role of community wardens in policing the "ASBO Generation"', *Urban Studies*, 50(3): 538–55.

Bunt, L. and Harris, M. (2010) *Radical Scotland: confronting the challenges facing Scotland's public services.* NESTA, London.

Children's Hearings Scotland (2015) *About Children's Hearings Scotland*, at: www. chscotland.gov.uk/ (accessed 29 July 2015).

Clayton, J., Donovan, C. and Merchant, J. (2015) 'Emotions of austerity care and commitment in public service delivery in the north east of England', *Emotion, Space and Society*, 14: 24–32.

Conradson, D. (2003) 'Doing organisational space: practices of voluntary welfare in the city', *Environment and Planning A*, 35(11): 1975–92.

Croall, H. (2006) 'Criminal justice in post-devolutionary Scotland', *Critical Social Policy*, 26(3): 587–607.

Dickens, L. and Lonie, D. (2013) 'Rap, rhythm and recognition: lyrical practices and the politics of voice on a community music project for young people experiencing challenging circumstances', *Emotion, Space and Society*, 9: 59–71.

Du Bois, D., Holloway, B., Valentine, J. and Cooper, H. (2002) 'Effectiveness of mentoring programmes for youth: a meta-analytic review', *American Journal of Community Psychology*, 30(2): 157–97.

Education Scotland (with Scottish Government and Youth Link Scotland) (2014) *National youth work strategy 2014–19*. Scottish Government, Edinburgh.

Evans, R. (2012) 'Sibling caringscapes: time-space practices of caring within youth-headed households in Tanzania and Uganda', *Geoforum*, 43(4): 824–35.

Gibson-Graham, J.K. (2008) 'Diverse economies: performative practice for "other worlds"', *Progress in Human Geography*, 32: 613–32.

Greenwood, P. (2008) 'Prevention and intervention programs for juvenile offenders', *Future of Children*, 18(2): 185–210.

Griffiths, M. (2013) 'The affective spaces of Big Society', paper presented at International and Interdisciplinary Conference on Emotional Geographies, Groningen, July 2013.

Grossman, J.B. and Rhodes, J.E. (2002) 'The test of time: predictors and effects of duration in youth mentoring relationships', *American Journal of Community Psychology*, 30(2): 199–219.

Hamnett, C. (2014) 'Shrinking the welfare state: the structure, geography and impact of British government benefit cuts', *Transactions of the Institute of British Geographers*, 39(4): 490–503.

Hopkins, P. and Alexander, C. (2010) 'Politics, mobility and nationhood: upscaling young people's geographies', *Area*, 42(2): 142–44.

Horton, J. and Kraftl, P. (2005) 'Editorial. For more-than-usefulness: six overlapping points about Children's Geographies', *Children's Geographies*, 3(2): 131–43.

Hunter, S. (2012) 'Ordering differentiation: reconfiguring governance as relational politics', *Journal of Psycho-Social Studies*, 6(1): 3–29.

Johnsen, S., Cloke, P. and May, J. (2005) 'Transitory spaces of care: serving homeless people on the street', *Health and Place*, 11: 323–36.

Judge, R. (2015) 'Emotion, volunteer-tourism and marginalized youth', in Blazek, M. and Kraftl, P. (eds) *Children's emotions in policy and practice: mapping and making spaces of childhood*. Palgrave, London, pp. 157–73.

Jupp, E. (2008) 'The feeling of participation: everyday spaces and urban change', *Geoforum*, 39(1): 331–43.

Jupp, E. (2013) '"I feel more at home here than in my own community": approaching the emotional geographies of neighbourhood policy', *Critical Social Policy*, 33(3): 532–53.

Keller, T. (2005) 'A systemic model of the youth mentoring intervention', *Journal of Primary Prevention*, 26(2): 169–88.

Korf, B. (2007) 'Antinomies of generosity: moral geographies and post-tsunami aid in Southeast Asia', *Geoforum*, 38(2): 366–78.

Kraftl, P. and Blazek, M. (2015) 'Mapping and making spaces of childhood', in Blazek, M. and Kraftl, P. (eds) *Children's emotions in policy and practice: mapping and making spaces of childhood*. Palgrave, London, pp. 308–11.

Laurie, N. and Bondi, L. (2005) *Working the space of neoliberalism*. Blackwell, Oxford.

Law, A. and Mooney, G. (2012) 'Devolution in a "stateless nation": nation-building and social policy in Scotland', *Social Policy and Administration*, 46(2): 161–77.

McAra, L. and McVie, S. (2010) 'Youth crime and justice: key messages from the Edinburgh Study of Youth Transitions and Crime', *Criminology and Criminal Justice*, 10(2): 179–209.

Meth, P. (2008) 'Vusi Majola – "Walking until the shoe is finished"', in Jeffrey, C. and Dyson, J. (eds) *Telling young lives: portraits of global youth.* Temple University Press, Philadelphia, PA, pp. 40–55.

Mills, S. and Kraftl, P. (eds) (2014) *Informal education, childhood and youth: geographies, histories, practices.* Palgrave Macmillan, London.

Moodie, M. and Fisher, J. (2009) 'Are youth mentoring programs good value-for-money? And evaluation of the Big Brothers Big Sisters Melbourne Program', *BMC Public Health*, 9(41), DOI: 10.1186/1471-2458-9-41.

Philip, K., King, C. and Shucksmith, J. (2004) *Sharing a laugh? A qualitative study of mentoring interventions with young people.* Joseph Rowntree Foundation, York.

Pryce, J. (2012) 'Mentor attunement: an approach to successful school-based mentoring relationships', *Child and Adolescent Social Work Journal*, DOI 10.1007/s10560-012-0269-6.

Punch, S. and Sugden, F. (2013) 'Work, education and out-migration among children and youth in upland Asia: changing patterns of labour and ecological knowledge in an era of globalisation', *Local Environment*, 18(3): 255–70.

Pykett, J. (2014) 'Representing attitudes to welfare dependency: relational geographies of welfare', *Sociological Research Online*, 19(3): 23.

Rhodes, J. and Lowe, S. (2008) 'Youth mentoring and resilience: implications for practice', *Child Care in Practice*, 14(1): 9–17.

Roberts, J.M. and Devine, F. (2004) 'Some everyday experiences of voluntarism: social capital, pleasure and the contingency of participation', *Social Politics*, 11(2): 280–96.

Sandford, R., Armour, K. and Stanton, D. (2010) 'Volunteer mentors as informal educators in a youth physical activity program', *Mentoring and Tutoring: Partnership in Learning*, 18(2): 135–53.

Schliehe, A.K. (2014) 'Inside "the carceral": girls and young women in the Scottish criminal justice system', *Scottish Geographical Journal*, 130(2): 71–85.

Scotland's Commissioner for Children and Young People (2014) *Children and Young People (Scotland) Bill: Briefing ahead of Stage 3 Debate, 19 February 2014*, at: www.sccyp.org.uk/policy/current-work (accessed 27 April 2014).

Scottish Government (2008a) *Getting it right for every child.* Scottish Government, Edinburgh.

Scottish Government (2008b) *Preventing offending by young people: a framework for action.* Scottish Government, Edinburgh.

Scottish Government (2009) *Multi-agency early and effective intervention: implementation guidance.* Scottish Government, Edinburgh.

Scottish Government (2013) *National youth justice practice guidance.* Scottish Government, Edinburgh.

Scottish Parliament (2014) *Children and Young People (Scotland) Bill.* Scottish Parliament, Edinburgh, at: www.scottish.parliament.uk/parliamentarybusiness/Bills/62233.aspx (accessed 27 April 2014).

Sercombe, H. (2009) *Young men in Scotland: a conversation.* YMCA Scotland, Edinburgh.

Smith, F.M., Timbrell, H., Woolvin, M., Muirhead, S. and Fyfe, N. (2010) 'Enlivened geographies of volunteering: situated, embodied and emotional practices of voluntary action', *Scottish Geographical Journal*, 126(4): 258–74.

Staeheli, L., Ehrkamp, P., Leitner, H. and Nagel, C. (2012) 'Dreaming the ordinary: daily life and the complex geographies of citizenship', *Progress in Human Geography*, 36(5): 628–44.

Tiffany, G. (2012) *'Positive for Youth': thoughts from a detached youth work point of view*, at: www.graemetiffany.co.uk/?p=604 (accessed 15 April 2014).

van Blerk, L. and Kesby, M. (eds) (2009) *Doing children's geographies: methodological issues in research with young people.* Routledge, London.

Wiles, J. (2011) 'Reflections on being a recipient of care: vexing the concept of vulnerability', *Social and Cultural Geography*, 12(6): 573–88.

Williamson, H. (2009) 'Integrated or targeted youth support services: an essay on "prevention"', in Blyth, M. and Soloman, E. (eds) *Prevention and youth crime: is early intervention working?* Policy Press, Bristol, pp. 11–21.

Afterword

Looking beyond our emotional present

Elizabeth A. Gagen

> Britain's stiff upper lip is no more. A nation once renowned for 'keeping it all in'
> is now determined to let it all out.
>
> (Doward, *The Observer*, 10 October 2004)

Introduction

It has been over ten years now since a survey entitled 'The Age of Therapy' de-
clared that the British public are now a 'feeling-friendly' nation, actively searching
out the expertise of therapists and counsellors in an attempt to solve problems
defined through emotions (Future Foundation, 2004). But while the development
of an emotional lexicon through which we process and value subjectivity is now
axiomatic, this volume of essays on emotional states demonstrates that there is still
a great deal of work to be done in unravelling the political, social and economic
power lines which run through this complex field. The preceding chapters offer a
timely and revealing survey of the myriad ways in which emotions have become so
much more than a constitutive part of ourselves, but have been co-opted into, and
become constitutive of, how we are governed. By bringing the focus on emotions
to bear on policy and practice, the volume builds on existing work in geography
and social science which, over the past two decades, has developed both a circum-
spect critique of the way emotions have infiltrated public life (Nolan, 1998; Furedi,
2004; Ecclestone and Hayes, 2009) and a broad appeal to better understand the
way emotions compose our very being in the world (Smith et al., 2009).

   Much of the recent work on emotions rests on the observation that emotions
have developed a new currency in public life. While there is significant evidence
to support this, the seismic nature of the shift is sometimes overestimated. In the
article from which the opening quote is taken, the writer declares that the 'British
have lost their reserve and are pouring their emotions out in therapy and counselling sessions' (Doward, 2004). Citing an earlier colonial period during which
emotions were suppressed and conduct was managed by hyper-vigilant discretion,
the article confirms a popular mythology that British emotions remained largely
hidden until the death of the Princess of Wales in September 1997. Often seen as
a cultural watershed in British history, Diana's death is frequently described as a
cathartic moment after which the British public gave themselves permission to

demonstrate a range of emotions previously confined to the private sphere (Kear and Steinberg, 1999). Taken alongside the supporting evidence that emotions are now central to education, business, government, urban planning, social and environmental policy, media and advertising, the overriding impression is that we are living in a new emotional era. In reflecting on the emotional state of contemporary social policy, however, I turn to Foucault's genealogical approach, which, rather than assuming a disconnect with the past, can better make sense of our present by recourse to the past (Foucault, 1984). Rather than limiting ourselves to the present, what more can be understood from enquiring about the practices, institutions and sets of knowledge that were necessary in order for our present to emerge in its current form? This method of constructing a history of the present does not mean dredging up historical precursors in order to demonstrate that current practice can be somehow linked to embryonic events or examples from the past (Garland, 2014). Rather, the focus of Foucault's genealogy is on discontinuity while at the same time finding historical conditions of existence upon which the present relies. I begin this Afterword by briefly reflecting on some of those historical entanglements as a reminder that, while our cultural and academic attention on emotions might have been brought into focus in recent years, that attention depends on conditions of possibility which were slower in gestation. I conclude by returning to the question of our current preoccupation with emotions and ask: what can be rescued from emotion work? There is much to debate and critique here, and indeed this volume gives us a great deal to consider critically. But what might we want to work with, embrace and develop into a set of practices that offer a more hopeful way forward?

Preconditions for the present

> Important and even invaluable political effects can be produced by historical analyses ... The problem is to let knowledge of the past work on the experience of the present.
>
> (Foucault, cited in Garland, 2014: 373)

There is ample evidence in this volume which suggests that the manipulation of affective states, the role of emotions in political subjectivities and the cultivation of valued emotional states in personal and professional life have gathered pace. But where does this current attention derive from? What institutions, truth regimes and political expertise were necessary to support our revaluing of emotions as a central condition of subjectivity and governance? And how does this help us better understand the current rationale of emotional governance? Nikolas Rose (1998, 1999) has written widely on the role of psychology in producing an epistemological field in which our subjectivity, our very sense of ourselves as subjects, was slowly transformed from a fundamentally private state to a calculable public resource. Over the course of the twentieth century, Rose (1998, 1999) demonstrates that the psychological sciences have been responsible for constructing the modern subject as autonomous, free-thinking and self-reflexive, capable of both choice and change. Across a huge range of enterprises, including government, education, human relations, war, parenting and business, the modern subject that we recognise as legitimate has been composed through truth claims fundamentally derived from psychological knowledge. This body of knowledge, more than any other, has provided the epistemological field through which we understand ourselves as subjects. While the emotional dimension of the psychological self is not necessarily front and central in Rose's analysis, it is nevertheless at its core. Through developments in clinical, occupational and educational psychology, social work, personnel, prison services, counselling and psychotherapy:

> Our thought worlds have been reconstructed, our ways of thinking about and talking about our personal feelings, our secret hopes, our ambitions, and disappointments. Our techniques for managing emotions have been reshaped. Our very sense of ourselves has been revolutionised. We have become intensely subjective beings.
>
> (Rose, 1999: 3)

Among all the technologies responsible for reimagining the self, psychotherapeutic principles have perhaps been the most powerful. The capacity for self-inspection and insight became dramatically revalued over the course of the twentieth century as part of a move away from blind obedience to a form of freedom that was predicated on self-understanding.

> The existence of a space of regulated freedom depended upon the generalisation of a set of ethical techniques for self-inspection and self-evaluation in relation to the code, a way of making the feelings, wishes, and emotions of the self visible to itself, a way in which citizens were to problematise and govern their lives and conduct, to find a way in which, as free subjects, they could live a good life as the consequence of their own character.
>
> (Rose, 1999: 228)

This form of regulated freedom which is capable of being governed was at the centre of the rise of the welfare state. The citizen with needs formed a contract with the government which provided social security and child welfare. Self-regulation was not only for the benefit of the individual and the qualifying currency of welfare, but was also for the greater social good. According to Rose (1998, 1999), emotional regulation has evolved symbiotically with government for much of the twentieth century. In thinking through the ways in which our current feeling-based forms of governance have taken new strides forward, what linkages and discontinuities can be found between then and now? If we think about these new intensities in emotional governance as part of this broader episteme, then how might our present emerge from this past?

One of the most obvious shifts in the late twentieth and early twenty-first century has been in the rise of brain science and its impact on modern conceptions of subjectivity. Much has been made of this neural turn, heralding a new era of 'brain culture' in which our cultural identities are fundamentally reshaped by neuroscientific knowledge and practice (Thornton, 2011; Pykett, 2015). However, following a genealogical commitment to considering the contingency of this present, Rose and Abi-Rached (2013) argue that neuroscience is not replacing previous psychological models of consciousness, but rather its knowledge is mapped onto these pre-existing sites.

> Neurobiological conceptions of personhood are not effacing other conceptions of who we are as human beings, notably those derived from psychology. On the contrary, they have latched on to them in the many sites and practices that were colonised by psychology across the twentieth century – from child rearing to marketing, and transformed them in significant ways.
>
> (Rose and Abi-Rached, 2013: 9)

Governing emotions is therefore no longer a contract with a psychologically self-reflexive citizen, but is emboldened by new knowledge claims which rely on neuroscientific materiality (Gagen, 2015).

Our contemporary governments have at their disposal new forms of knowledge and expertise derived from neuroscience which form the basis of new ways of governing. One area of governance which has received close attention is the behaviour change agenda which draws on a range of recent publications by leading neuroscientists that identify the emotional basis of much human behaviour (Jones et al., 2013). In an attempt to pre-empt emotionally driven decisions and encourage actions that are in the interests of the wider community and individual wellbeing, governments have developed strategies to put in place default options that encourage 'responsible' decisions (ibid.). The development of behaviour change policies relies on a paradigmatic current in neuroscientific thinking. Traditionally brain science locates the centre of decisions in the neocortex where cognition was thought to take place, but since the mid-1990s this was replaced by the theory that the amygdala – responsible for emotional reactions – controlled far more behavioural decisions than was previously imagined (see Damasio, 1994, 1999; Le Doux, 1996, 2002). It was this knowledge that persuaded governments to rethink traditional forms of persuasion and control. If neuroscience was correct and human behaviour is governed principally through emotional and instinctive responses, then providing citizens with a set of choices which pre-determine socially responsible outcomes becomes the best way to govern (Jones et al., 2013).

The reach of neuroscientific knowledge claims has been inestimable, shaping social, economic, environmental and education policy in profound ways. Indeed, most of the empirical sites and cases in this volume are, in some way, informed by these recent changes. To take these patterns of governing as the beginning and end point of analysis, however, not only limits the analysis to the present but risks conveying a misplaced sense of rupture with the past. Far better, to echo Foucault, would be to use this present as part of a diagnosis of a problem (Dreyfus and Rabinow, 1982). How do our current emotionally attuned forms of governance continue or rework established goals of governmentality? Where do these current themes of empathy, care and affect emerge from? How is the government of subjectivity maintained *despite* a shift in the methods and targets of reform?

As a preliminary attempt at such an analysis, I turn now to two vignettes, one recent and, no doubt, familiar, the other from a more distant location and, perhaps, less familiar. Both, I argue, are epistemically connected to the present emotional era, suggesting that governing through emotions is not new but newly vitalised by a critical mass of knowledge claims that have provided governments with the opportunity to govern differently.

Instincts and impulses

> It is not easy to realize that it was only a single generation ago when we used to think that the animals are ruled by instinct, man by reason. We know better now. What was once the 'new' psychology has taught us that man has more separate instincts than any other creature that breathes, and that however superior his rational life, it is still based upon a substructure of primitive instincts which he shares with the beasts of the field.
>
> (J. Adams Puffer, 1912, quoted by Schlossman, 1973: 144)

At the heart of the current behaviour change agenda and central to recent attempts in education policy to teach children better impulse control is a deep anxiety about the powerful nature of human impulse (Jones et al., 2013; Gagen, 2015). The emotional nature of decision-making, the unthinking actions that drive impulsive decisions, revealed by neuroscientific investigations as a fundamental part of the human fabric, are seen by policy-makers to cause harm to both individuals and society. Policy interventions are designed to both limit choice and equip individuals with the skills to control their own emotional nerve centres. But where did this preoccupation emerge from and what sets of knowledge were used to exercise power prior to this emotional turn? I begin by returning to an example of measurement and intervention in education policy at the turn of the twentieth century.

Governing embodied impulses

During the 1890s, the scientific measurement of child development emerged as a widespread practice across the United States and Europe. In February 1899, the Committee on Physical Culture of Chicago Public Schools agreed to appoint two child study experts to carry out anthropomorphic studies on elementary school children throughout the city. Mr Fred W. Smedley and Mr C. Victor Campbell were appointed to carry out the studies which took place between March and June 1899. The Child Study Report describes the range of data collected, including height, weight, age, gender, lung capacity and eyesight. In addition, some more specific physical and mental tests were carried out, the most important of which used the ergograph. Invented in 1890 by the Italian physiologist Angelo Mosso, it recorded the force and frequency of finger flexion and was considered to be a proxy measurement of strength, muscle control and, by extension, self-control.

In order to test children's ability to exercise controlled muscle movements, the child's forearm was clamped into the ergograph, a cord travelled from the child's middle finger to a pulley loaded with a weight. As the child flexed the finger, the weight lifted and, simultaneously, a pen placed a mark on the kymograph. The child was asked to repeat this motion, in time with a metronome, for 90 seconds. For each child, their individual ergogram – the tracing produced by the finger movements – was taken as a pronouncement of their nervous state. A well-controlled child would produce an ergogram of evenly spaced movements which kept time with the metronome; a nervous child would produce erratic nerve impulses, sometimes slower, sometimes faster than the metronome. 'The want of muscular control is shown by the irregularity in the heights of the lifts, the uneven spacing and the excessive number of lifts, showing the failure to keep time with the metronome' (Board of Education of Chicago, 1899–1900). Ergograph measurements were taken across urban school systems in America during this period in an attempt to monitor and record physical and mental health capacity, with the ergograph standing out as 'probably the best means to detect the nervous condition of the school children' (Joint Committee of the Women's Clubs of Pittsburgh and Vicinity, 1900, p. 62).

The use of the term impulse is quite specific here, and different from the way current policy-makers recognise impulse control as a feature of behaviour. However, I suggest that there are, if not similarities, then connections to modern forms of governing subjectivity, specifically the use of the ergogram as a proxy for self-control. At this time, as child psychology was just emerging as a coherent field, morality, intelligence and character were seen as inherently measurable via physical movement and performance (Gagen, 2004, 2006). Within this analytic field, it was entirely logical to extrapolate conclusions regarding a subject's governability from an experiment in rhythmic exercise. As the reports from Chicago state: 'It is clear from the foregoing charts and tables, that on average those pupils who have made great intellectual advancement are on the whole taller, heavier, stronger, possessed of greater endurance, and larger breathing capacity than those who have made less advancement' (Board of Education of Chicago, 1899–1900). The control of muscles was a rewarding feature of subjectivity and therefore justified government interventions in physical health. The results from the child study were used to mount a campaign for comprehensive physical exercise regimes in schools across urban America as a way of improving self-control and producing more governable citizens.

Governing impulsive behaviour

Over half a century later, after psychology had established itself as a formal scientific discipline, another, now infamous, experiment took place. Known colloquially as the 'marshmallow test', Walter Mischel's experiment into the exercise of delayed gratification in children has become one of the central pillars of modern emotional intelligence literature, a common reference point in popular neuroscience, and a well-used citation in behaviour change policy (Gagen, 2015). The experiments took place between 1968 and 1972 at the Bing Nursery School, a purpose-built laboratory established a few years earlier by the Stanford University Psychology Department (Lehrer, 2009). In the experiment, groups of children were invited into a room. In the room were desks and chairs. At the front was a tray of treats. Each child could choose from a pretzel, a marshmallow or a cookie. Most children chose the marshmallow. They were instructed to take their treat to their assigned desk and told that they could choose to eat their treat immediately. Alternatively, if they could wait an unspecified length of time while the researcher left the room, they would receive two treats instead of one. If they could not wait and opted for the one immediate treat, they were asked to ring a bell to indicate their decision. The experiments were video recorded to verify the point at which a child 'broke' and ate the treat, and to observe the various strategies children used to resist the impulse to seek the immediate but lesser gratification of one marshmallow versus two.

The goal of the experiment was to understand how some individuals were able to control their desires and defer gratification, while others were not. However, as the experiments continued, Mischel began to speculate about the larger implications of the study, and the discrete experiment expanded into a longitudinal study of the life outcomes of test subjects. In 1981, Mischel began gathering data on the progress of the children who had taken part in the original tests, by then in high school and approaching graduation. He found that the subjects who had been unable to wait the time necessary to double the marshmallow prize had, on average, lower SAT scores, a higher incidence of behavioural problems, a lower attention span, and they struggled with stressful situations and found it harder to make friends. And so impulse control began to assume great importance as a feature of behaviour and broader life outcomes. Not only was Mischel able to demonstrate that later performance was strongly correlated with the ability to delay gratification, but through his video footage he also claimed to offer insight into strategies which could teach individuals better self-control. Referred to as the 'strategic allocation of attention', similar in vein to the idea of metacognition, Mischel argued that the children who evidenced self-control did so by distracting themselves from the object of desire (Lehrer, 2009). He designed a series of mental tricks involving visualisation that children could use to distract themselves from their desire.

In the years that followed, Mischel's experiment was taken up by proponents of what became known as emotional intelligence (Goleman, 1995). Armed with emerging evidence from neuroscience that the emotional brain is central to identity and behaviour, the role of impulse control, in particular, came to be regarded as a key feature of individual success (Gagen, 2015). From the ability to defer television viewing in favour of homework, to save for a pension rather than spend on a holiday, to consider a healthy meal rather than grab an immediate snack, to refrain from hot-headed arguments in favour of self-calming techniques; all became newly regarded as praiseworthy features of the emotionally intelligent subject. Impulse control, like muscle control at the end of the previous century, developed into a proxy for emotional intelligence, which itself has become the gold standard of governable subjectivity. Rather than focus on the bodily control of individual subjects, therefore, governmentality has shifted to its current behavioural preoccupation. In thinking through the questions which drive this volume, questions that ask why governments have come to value feelings, what new modes of government this engenders, and who might be excluded from this regime, I would add: what kind of individual subject is necessary for emotional forms of governance to be successful, and from where did this subjectivity derive?

Conclusions

There is no doubt that the changes brought about by neuroscience, behavioural economics and cognitive psychology offer governments new ways of governing through emotions. This volume makes a valuable and necessary step towards understanding how and where this takes place, and begins to unpick the various power lines that cross-cut these operations in all their heterogeneous sites. I have sought to add to this discussion by suggesting that still more can be gained from using the present context as a springboard for understanding its emergence from practices, institutions and knowledge claims in the past. Inherent in this approach, however, is a commitment to looking forward. As Garland (2014: 372, emphasis added) notes, a commitment to tracing the history of the present involves tracing 'an often aleatory path of descent and emergence that suggests the contingency of the present *and* the openness of the future'. What, then, might we claim back from current practices of emotional governance that promise a more humane and compassionate future? Are there aspects of emotional governance that offer us the possibility of living in a social world that acknowledges the emotional contingency of our situated lives? Are there ways of understanding emotions that might genuinely help the practices of governance happen more humanely? Some work has already begun in this direction (see Pykett et al., Chapter 1, and Jupp, Chapter 10, in this volume). But to make more assertive advances towards this goal, we need to think more openly and thoughtfully about what we can learn from practices like psychotherapy, phenomenology and psychoanalysis, practices which are alive to the vital, relational, embodied encounters which comprise our emotional lives (Bondi, 2005, 2014). It is ten years now since Bondi (2005) argued that, despite emotional geographies generating significant interest, there is a tendency to consider emotions through conceptual frameworks that limit the radical potential of emotional engagement. Rather, she argued, there is a tendency to focus on 'wider cultural trends that treat emotions as individualised attributes available for commercial and political exploitation' (Bondi, 2005: 445). I recognise myself here. And while I acknowledge the merit of researching the cultural value of emotions, and have devoted considerable energy to this form of analysis, I accept its limits. In thinking forward as well as back, we need to take seriously emotional geographies' call to develop an understanding of subjectivity that is not just strategic but acknowledges the 'diffuse and all pervasive and yet also heart and gut-wrenchingly present and personal' nature of emotional constitution (Smith et al., 2009: 3).

References

Board of Education of Chicago (1899–1900) Forty-Sixth Annual Report: Child Study Report No. 2, at: https://archive.org/stream/reportchildstudy00chic/reportchild study00chic_djvu.txt (accessed 24 August 2015).

Bondi, L. (2005) 'Making connections and thinking through emotions: between geography and psychotherapy', *Transactions of the Institute of British Geographers*, 30: 433–48.

Bondi, L. (2014) 'Understanding feelings: engaging with unconscious communication and embodied knowledge', *Emotion, Space and Society*, 10: 44–54.

Damasio, A. (1994) *Descartes' error*. Putman, New York.

Damasio, A. (1999) *The feeling of what happens: body, emotion and the making of consciousness*. Harcourt Brace, New York.

Doward, J. (2004) 'How we fell in love with the couch', *The Observer*, 10 October 2004, p. 3.

Dreyfus, H. and Rabinow, P. (1982) *Michel Foucault: beyond structuralism and hermeneutics*. University of Chicago Press, Chicago, IL.

Ecclestone, K. and Hayes, D. (2009) *The dangerous rise of therapeutic education*. Routledge, London.

Foucault, M. (1984) 'Nietzsche, genealogy, history', in Rabinow, P. (ed.) *The Foucault reader*. Pantheon, New York, pp. 76–100.

Furedi, F. (2004) *Therapy culture: cultivating vulnerability in an uncertain age*. Routledge, London.

Future Foundation (2004) *The age of therapy: exploring attitudes towards and acceptance of counselling and psychotherapy*. The Future Foundation, London.

Gagen, E.A. (2004) 'Making America flesh: physicality and nationhood in early twentieth-century physical education reform', *Cultural Geographies*, 11(4): 417–42.

Gagen, E.A. (2006) 'Measuring the soul: psychological technologies and the production of physical health in Progressive Era America', *Environment and Planning D*, 24(6): 827–49.

Gagen, E.A. (2015) 'Governing emotions: citizenship, neuroscience and the education of youth', *Transactions of the Institute of British Geographers*, 40(1): 140–52.

Garland, D. (2014) 'What is a "history of the present"? On Foucault's genealogies and their critical preconditions', *Punishment and Society*, 16(4): 365–84.

Goleman, D. (1995) *Emotional intelligence: why it can matter more than IQ*. Bloomsbury, London.

Joint Committee of the Women's Clubs of Pittsburgh and Vicinity (1900) Reports of the Joint Committee, Pittsburgh Playground Association 1897–1908, Carnegie Library.

Jones, R., Pykett, J. and Whitehead, M. (2013) *Changing behaviours: on the rise of the psychological state*. Edward Elgar, Cheltenham.

Kear, A. and Steinberg, D.L. (1999) *Mourning Diana: nation, culture and the performance of grief*. Routledge, London.

LeDoux, J. (1996) *The emotional brain*. Simon and Schuster, New York.

LeDoux, J. (2002) *Synaptic self: how our brains become who we are*. Penguin, London.

Lehrer, J. (2009) 'Don't! The secret of self-control', *New Yorker*, 18 May 2009, at: www. newyorker.com/magazine/2009/05/18/dont-2 (4 September 2015).

Nolan, J.L. (1998) *The therapeutic state: justifying government at century's end.* New York University Press, New York.

Pykett, J. (2015) *Brain culture: shaping policy through neuroscience.* Policy Press, Bristol.

Rose, N. (1998) *Inventing ourselves: psychology, power and personhood.* Cambridge University Press, Cambridge.

Rose, N. (1999, second edition) *Governing the soul: the shaping of the private self.* Free Association Books, London.

Rose, N. and Abi-Rached, J.M. (2013) *Neuro: the new brain sciences and the management of the mind.* Princeton University Press, Princeton, NJ.

Schlossman, S.L. (1973) 'G. Stanley Hall and the boys' club: conservative applications of recapitulation theory', *Journal of the History of Behavioural Sciences*, 9(2): 140–47.

Smith, M., Davidson, J., Cameron, L. and Bondi, L. (2009) 'Geography and emotion – emerging constellations', in Smith, M., Davidson, J. and Cameron, L. (eds) *Emotion, place and culture.* Ashgate, Farnham, pp. 1–18.

Thornton, D.J. (2011) *Brain culture: neuroscience and popular media.* Rutgers University Press, Piscataway, NJ.

Index